AF588051

Synthesis Lectures on Sustainable Development

Series Editor

This series publishes short books related to sustainable development practices relevant to engineers, technologists, managers, educators, and policy makers. The books are organized around the United Nations Sustainable Development Goals for 2015–2030. Design for sustainability along with the economics of sustainable development will be the common themes for each book. Topics to be covered will span all the major engineering disciplines along with the natural and environmental sciences that contribute to an understanding of sustainable development. The goal of this series is to make theory and research accessible to practitioners working on sustainable development efforts.

Navnath T. Hatvate · Ajinkya Madan Satdive ·
Hemantkumar N. Akolkar · A. K. Haghi

Plastic Waste Management

Solutions for Sustainable Development

Navnath T. Hatvate
Institute of Chemical Technology, Mumbai,
Marathwada Campus
Jalna, Maharashtra, India

Ajinkya Madan Satdive
Institute of Chemical Technology, Mumbai,
Marathwada Campus
Jalna, Maharashtra, India

Hemantkumar N. Akolkar
Department of Chemistry
Rayat Shikshan Sanstha's Abasaheb Marathe
Arts and New Commerce Science College
Rajapur, Maharashtra, India

A. K. Haghi
Department of Chemistry
Institute of Molecular Sciences
University of Coimbra
Coimbra, Portugal

ISSN 2637-7675 ISSN 2637-7691 (electronic)
Synthesis Lectures on Sustainable Development
ISBN 978-3-031-96659-0 ISBN 978-3-031-96660-6 (eBook)
https://doi.org/10.1007/978-3-031-96660-6

This Springer imprint is published by the registered company Springer Nature Switzerland AG
The registered company address is: Gewerbestrasse 11, 6330 Cham, Switzerland

Preface

A new global catastrophe has emerged as a result of the accumulation of trash made of plastic. Plastic pollution's effects on ecosystems, human health, and the economy are urgent issues that demand quick attention and coordinated action. With the title *Plastic Waste Management—Solutions for Sustainable Development*, this book seeks to give readers a thorough understanding of the problem of plastic waste management, its effects on the environment and human health, and possible solutions for a sustainable future.

From the varieties of plastic materials and their degradation patterns to the effects of plastic pollution on the environment and human health, the book's four chapters examine the challenges of managing plastic trash. The chapters also look at current methods for managing plastic garbage, stressing their shortcomings and difficulties, and talk about fresh ideas and different strategies for doing so.

Researchers, legislators, business leaders, and anybody else interested in comprehending the problem of plastic waste management and investigating options for a sustainable future are the target audience for this book. It is our aim that this book will add to the current conversation on plastic pollution and motivate action to lessen its effects.

In addition to inspiring fresh concepts and inventions to advance a more sustainable future, we believe that this book will prove to be an invaluable tool for everyone attempting to comprehend and tackle the problem of plastic waste management.

Jalna, India — Navnath T. Hatvate
Jalna, India — Ajinkya Madan Satdive
Rajapur, India — Hemantkumar N. Akolkar
Coimbra, Portugal — A. K. Haghi

Contents

1 Introduction 1
1.1 Plastics and the Environmental Hazards 1
1.1.1 Impact of Plastics on Land 1
1.1.2 Impact of Plastics on Aquatic Life 2
1.1.3 Impact of Plastic on Air Pollution 4
1.1.4 Impact of Plastic on Water Pollution 4
1.2 Plastics, the Environment and Human Health 5
1.2.1 Bisphenol-A (BPA) 6
1.2.2 Phthalates 9
1.2.3 Polychlorinated Biphenyls (PCBs) 9
1.2.4 Halogenated Flame Retardants 10
1.2.5 Emissions from Plastic Combustion 11
1.3 Plastic Waste Pollution and Sustainable Solutions 12
1.3.1 Reduction and Prevention Strategies 13
1.4 Plastic Waste Management Strategies in Practice 16
1.4.1 Reduction and Prevention Strategies 17
1.4.2 Collection and Sorting Strategies 18
1.4.3 Recycling and Reuse Strategies 18
1.4.4 Waste-To-Energy Technologies 19
1.5 Conclusion 20
References 20

2 Current Solutions for Sustainable Ecosystems 31
2.1 Introduction 31
2.2 Major Challenges to Sustainable Ecosystems 31
2.2.1 Pollution 31
2.2.2 Climate Change and Global Warming 32
2.2.3 Land Degradation and Agricultural Constraints 33
2.2.4 Habitat and Biodiversity Loss 33

2.3 A Complex Issue with Complex Solution 33
2.4 Toxic Reduction of Plastic Waste 35
2.4.1 Sources and Types of Toxic Substances in Plastic Waste 35
2.4.2 Technologies and Methods for Reducing Toxicity in Plastic Waste Management 37
2.5 Efficiency of Current Plastic Waste Management 40
2.5.1 Assessment of Existing Plastic Waste Management Systems 40
2.5.2 Evaluation of Efficiency Metrics and Performance Indicators 41
2.5.3 Challenges and Limitations in Current Practices 41
2.5.4 Future Directions for Improving Plastic Waste Management 42
2.6 Plastic Recycling 42
2.6.1 Types and Recyclability of Plastics 44
2.6.2 Environmental Benefits of Plastic Recycling 44
2.6.3 Role of Stakeholders in Enhancing Recycling 44
2.7 Conclusion 46
References 47

3 Innovative Approaches to Plastic Waste Management 51
3.1 Introduction 51
3.2 Successful Local Experiences in Plastic Waste Management 52
3.2.1 Community-Led Plastic Waste Initiatives 52
3.2.2 Government-Led Policy Interventions 53
3.2.3 Corporate and Industrial Best Practices 56
3.2.4 Technological Innovations in Waste Management 57
3.3 Sustainable Disposal Solutions for Plastic Waste Management 59
3.4 Innovations in Waste Treatment Technologies 62
3.4.1 Advanced Recycling Technologies 63
3.4.2 Nanotechnology in Plastic Decomposition 63
3.4.3 Biotechnological Solutions 64
3.4.4 Automation in Waste Processing 64
3.5 Innovations in Plastic Pollution Control 65
3.5.1 Policy-Based Strategies 66
3.5.2 Ocean Clean-Up Technologies 67
3.5.3 Microplastic Filtration Solutions 68
3.6 Plastic in Aquatic Environments 69
3.6.1 Sources of Marine Pollution 70
3.6.2 Impact on Marine Life and Ecosystems 72
3.6.3 Global Cleanup and Prevention Efforts 74
3.7 Conclusion 76
References 76

4 Future Prospects for a Sustainable Environment 83
4.1 Introduction 83
4.2 Methods for Measuring Plastic Pollution 84
4.2.1 The Challenge of Microplastic Detection 84
4.2.2 Sampling and Collection Methods 84
4.2.3 Separation and Purification Techniques 85
4.3 Plastic Recycling Technologies: Research and Development 86
4.3.1 Mechanical Recycling 87
4.3.2 Chemical Recycling 87
4.3.3 Biological Recycling 89
4.4 Remediation of Plastic Waste 90
4.4.1 Physical Remediation Techniques 91
4.4.2 Chemical and Advanced Oxidation Processes (AOPs) 91
4.4.3 Bioremediation and Phytoremediation 92
4.4.4 Emerging Approaches 92
4.5 Circular Economy of Plastics 92
4.6 Policy Implications and Recommendations 94
4.6.1 Policy Shortcomings and the Need for Integration 94
4.6.2 System Dynamics Modeling 95
4.6.3 Global Policy Landscape 96
4.6.4 Policy Implications and Recommendations 96
4.7 Conclusion 98
References 99

About the Authors

Dr. Navnath T. Hatvate, Ph.D. is an Assistant Professor at the Institute of Chemical Technology, Mumbai, Marathwada Campus, in Jalna. He holds a Master of Pharmacy and a Ph.D. from the Department of Pharmaceutical Sciences and Technology at the same institute. With 11 years of research experience and 6 years of teaching experience, he specializes in several areas. His research interests include the development of new chemical entities (NCE) and their analogues, modifications of excipients and their applications in drug delivery, and the use of artificial intelligence (AI) and machine learning (ML) in drug discovery and delivery. He is also focused on the process chemistry of active pharmaceutical ingredients (APIs) and intermediates, total synthesis of natural products and bioactive compounds, and the application of novel technologies in organic synthesis. Dr. Hatvate has published 52 research papers and authored 4 books. Additionally, he has guided 18 master's students in their research projects.

Mr. Ajinkya Madan Satdive is currently an Assistant Professor at the Institute of Chemical Technology, Marathwada Off Campus in Jalna. He is a dedicated professional specializing in polymer engineering and technology, with a diverse background encompassing both industry and academia. With nearly 8 years of combined experience, his current focus is on the modification of biopolymers, material modification techniques, polymer processing methodologies, synthesis strategies, elucidating structure-property relationships, and the development of polymer blends and alloys. He has taught various subjects, including polymer processing, polymer blends and alloys, high polymer chemistry, and polymer science and technology. To date, he has published 8 international peer-reviewed articles and 6 book chapters.

Dr. Hemantkumar N. Akolkar, Ph.D. is currently working as an Associate Professor in the Department of Chemistry at Rayat Shikshan Sanstha's Abasaheb Marathe Arts and New Commerce Science College, Maharashtra, India. He is an acclaimed academician and researcher who has worked for the last 14 years at both undergraduate and postgraduate levels. He has published 55+ research papers in national and international journals and conferences of repute. His areas of interest are heterocyclic chemistry, synthetic organic chemistry, and green chemistry. He is a reviewer of several journals of international repute in Chemistry.

A. K. Haghi is a retired Professor and has written, co-written, edited, or co-edited more than 1000 publications, including books, book chapters, and papers in refereed journals with over 4200 citations and h-index of 34, according to the Google Scholar database. Prof. Haghi holds a B.Sc. in urban and environmental engineering from the University of North Carolina (USA) and holds two M.Sc. degrees, one in mechanical engineering from North Carolina State University (USA) and another one in applied mechanics, acoustics, and materials from the Université de Technologie de Compiègne (France). He was awarded a Ph.D. in engineering sciences at Université de Franche-Comté (France).

Professor Haghi's extensive educational background and supervisory roles underscore his expertise and contributions to the field of engineering sciences. He is appointed as Honorary Research Associate (HRA) at University of Coimbra, Portugal. He is a regular reviewer of leading international journals.

List of Figures

Fig. 1.1 Environmental impact pathways related to plastic waste management (*Created by using Biorender.com*) . . . 2
Fig. 1.2 Effect of plastic waste on human health (*Created by using Biorender.com*) . . . 6
Fig. 1.3 Annual plastic waste by disposal method (2000–2019) (*Created by using Biorender.com*) . . . 17
Fig. 2.1 Plastic recycling process (*Created by using Biorender.com*) . . . 37
Fig. 2.2 Comprehensive framework for tackling microplastic pollution (*Created by using Biorender.com*) . . . 40
Fig. 2.3 7R of recycling process (*Created by using Biorender.com*) . . . 43
Fig. 3.1 Plastic pyrolysis pathway for resource and energy recovery (*Created by using Biorender.com*) . . . 60
Fig. 3.2 Integrated chemical processes for plastic recycling and upcycling (*Created by using Biorender.com*) . . . 61
Fig. 3.3 Technological approaches to plastic waste management (*Created by using Biorender.com*) . . . 65
Fig. 4.1 Thermal degradation of plastic (*Created by using Biorender.com*) . . . 88
Fig. 4.2 Photodegradation mechanism of plastic (*Created by using Biorender.com*) . . . 93
Fig. 4.3 Microbial degradation of plastic (*Created by using Biorender.com*) . . . 93
Fig. 4.4 Circular economy of plastic waste (*Created by using Biorender.com*) . . . 95

List of Tables

Table 1.1 Chemicals and additives in plastic production-associated health effects and plastic types 7
Table 2.1 Types of pollution associated with plastic and their toxic byproducts 32
Table 2.2 Major components present in plastic waste 37
Table 4.1 Advantages and impacts of key innovations in plastic waste recycling and circular economy 90

1 Introduction

1.1 Plastics and the Environmental Hazards

The most widely used materials in society are plastics, which are mainly derived from petrochemicals due to their affordability, durability, and lightweight properties. After being used, plastic products are discarded into the environment as plastic waste. Because plastics are composed of carbon-based organic materials with long chains and cross-linking polymers, they are very difficult for the environment to break down, which poses a disposal issue. When low-density plastics are dumped into water bodies, they float on the water's surface and harm aquatic life. Certain chemicals employed in the manufacture of plastic have extremely harmful effects on the environment and living things. Its durability and toxic substances lead to several environmental effects, including air, water, and soil contamination (Kehinde et al., 2020). Figure 1.1 illustrates the various ecological impact pathways associated with plastic waste management.

1.1.1 Impact of Plastics on Land

A significant amount of plastic waste is produced every day, which may then be dumped in a landfill, incinerated, recycled, or released into the environment. This leads to environmental pollution. Pollution from plastic and plastic products has the potential to pollute and harm the terrestrial ecosystem, subsequently migrating to the aquatic environment. Hazardous compounds are released into the soil when plastic is burned or degraded abiotically and biotically (Zubris and Richards, 2005). Contaminated plastics can release toxic chemicals into soil, which later percolate into water sources in the surrounding area, which could later flow into underground water and other water sources. As a result, the soil deteriorates, and the organisms that consume this water are affected by disease.

N. T. Hatvate et al., *Plastic Waste Management*, Synthesis Lectures on Sustainable Development, https://doi.org/10.1007/978-3-031-96660-6_1

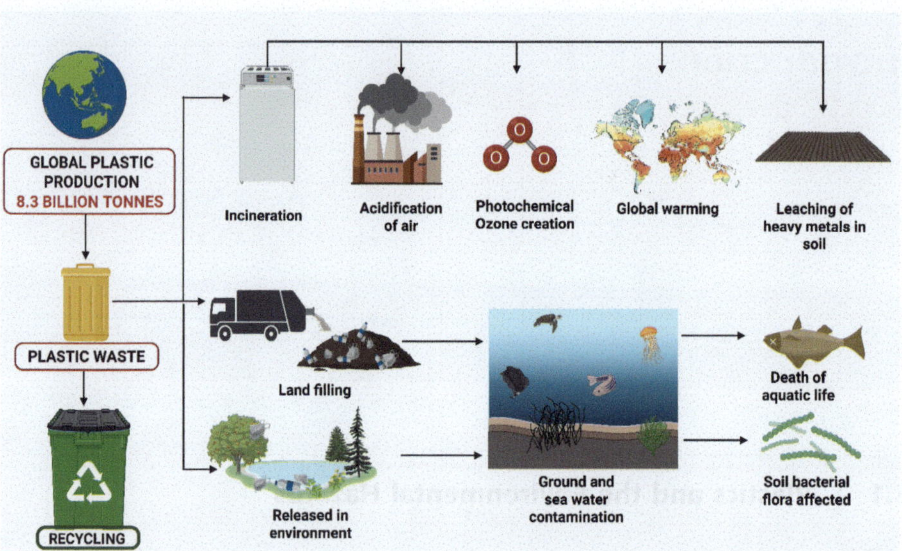

Fig. 1.1 Environmental impact pathways related to plastic waste management (*Created by using Biorender.com*)

Plastics are dumped in landfills, where they are continuously accumulating in significant amounts (Kehinde et al., 2020). This results in the alteration of the structure of the soil due to metals, including lead, cadmium and mercury, pharmaceutical waste, disinfectants and chemical substances (Hakeem et al., 2018). Stabilizers, plasticizers, and hazardous colorant moieties used as additives in plastic also leach and seep into the soil and water when plastics are disposed of in landfills (Zubris and Richards, 2005). The chlorinated plastics degrade and release certain harmful compounds that seep into the surrounding soil and aquatic systems. According to reports, landfills produce 20% of greenhouse gases (GHG), including carbon dioxide (CO_2), methane (CH_4), sulfur, and nitrogen gases from the decomposition of plastic. These gases can result in landfill fires and lower solar radiation (Kehinde et al., 2020). When plastic mulching is used in agriculture, it raises concerns about microplastic, which affects soil fertility, depletes nutrients, deteriorates soil health, and releases harmful chemicals like bisphenol A (BPA) into the soil. Crops are also contaminated by continuous chemical deposition into the soil (Islam et al., 2023).

1.1.2 Impact of Plastics on Aquatic Life

The world's ecosystem is made up of 70% water, and the majority of life on Earth depends on water sources, such as rivers, streams, and oceans, which supply the dissolved oxygen needed for breathing. Aquatic animals and other living things are concerned about waste made of plastic. It is estimated that 165 million tons of plastic trash were present in the world's oceans in 2012 and that at least 8 million tons of plastic waste will enter the ocean each year (Okunola et al., 2019a). Plastic waste enters the water from combined sewer overflow, groundwater run-off, littering and solid waste disposal. Plastic waste pollution primarily poses a concern to coastal areas because plastic particles can endure natural degradation processes for years in aquatic environments (Matavos-Aramyan, 2024). Additionally, plastic bottle waste and ship spills harm the oceans. As plastic sediments reached the seawater, a dangerous concentration of plastic debris developed. Pollutants such as organic compounds and heavy metals are formed when plastic waste is deposited into the ocean (Kehinde et al., 2020). Plastic can partially decompose in the sea within a year, but not entirely. Harmful chemicals, such as BPA and polystyrene, are released into oceans during the process of plastic breakdown (Eriksen et al., 2014). Additionally, the plastic waste is naturally nonpolar and repels water molecules, it releases biologically hazardous waste in the form of defoliants like dichlorodiphenyltrichloroethane (DDTs) and polychlorinated biphenyls (PCBs) (Kehinde et al., 2020). Plastic waste contains organic pollutants that harm aquatic and terrestrial life when released into the environment, such as vinyl chloride and dioxins (found in PVC), benzene (found in polystyrene), formaldehyde, phthalates, and other plasticizers (found in PVC and other materials), and bisphenol-A (found in polycarbonate, or BPA). Chlorinated plastic releases hazardous substances into surrounding soil that percolate into groundwater, which can damage species that drink the water (Proshad et al., 2017). The entire food chain that consumes seafood is contaminated by toxic pollutant chemicals that are released into the ocean by plastic debris and subsequently enter the body through the skin and tissue or are ingested by marine life. Aquatic organisms quickly colonize plastic waste that floats on the water surface and becomes persistent. Contaminants such as persistent organic pollutants-including nonylphenol, PCBs, DDE (dichlorodiphenyldichloroethylene), and phenanthrene can accumulate on plastic debris in concentrations many times higher than those found in the surrounding seawater (Okunola et al., 2019a).

Due to its small size and presence in both benthic and pelagic ecosystems, microplastic is bioavailable to aquatic life (Barboza and Gimenez, 2015). It is estimated that 83% of water sources are contaminated with plastic residue, and per year, more than 1 lakh aquatic animals die. Marine life, including fish, turtles, insects, and mammals, becomes entangled in or consume plastic waste, which can cause sores, cuts, and even death (Gregory, 2009). The presence of plastic waste in aquatic environments can hinder fish reproduction, eliminate beneficial species, and interfere with natural ecological processes (Proshad et al., 2017).

1.1.3 Impact of Plastic on Air Pollution

Plastics are a major source of air pollution due to their manufacture, burning, and deterioration. The manufacturing process contributes to smog and climate change by releasing greenhouse gases and volatile organic compounds (VOCs) (Jansen et al., 2024; Sohel Parvez et al., 2024). Volatile organic compounds (VOCs) generated during the synthesis of plastic can react with nitrogen oxides (Nox) to form ground-level ozone, which is considered harmful to plants and human beings (Seewoo et al., 2024; Villarrubia-Gómez et al., 2024). In addition, open burning of plastics is a prevalent practice in areas with poor waste management infrastructure, which mostly causes air pollution (Verma et al., 2016; Pathak et al., 2023). Among the dangerous pollutants released by this process include carbon monoxide, furans, dioxins, and polycyclic aromatic hydrocarbons (PAHs) (Bastian et al., 2013). Amongst the above, the highly hazardous persistent organic pollutants (POPs) dioxins and furans can accumulate in both human and animal bodies, causing significant health difficulties such as cancer, immune disorders, Chronic Obstructive Pulmonary Disorders (COPD), and hormonal changes (Velis and Cook, 2021; Zhang et al., 2017).

Microplastic air pollution is an increasingly important issue that requires our attention and action. Recent research shows that microplastics are airborne in urban and rural areas, indicating that they can travel from urban to rural areas (Yao et al., 2022; Vattanasit et al., 2023). The main sources of atmospheric microplastics are the breakage of bigger plastic waste, the abrasion of synthetic textiles, and tyre wear from automobiles (Wang et al., 2021). As discussed earlier, this microplastic is also inhaled, which may cause irritation to the mucosal membrane of the respiratory tract, which leads to inflammation and oxidative stress. Microplastics may contain toxic chemicals which enter into humans and the biosphere, including heavy metals and POPs (Carriera et al., 2024).

1.1.4 Impact of Plastic on Water Pollution

Water pollution by plastic is also one of the major challenges to human beings. For instance, The Ganges River, 2,500 km long, supports 655 million people in South Asia but is heavily polluted with plastic waste (Gupta et al., 2024; Nelms et al., 2021). The Ganges is the major source of drinking water, agricultural products, and also supports religious ceremonies, but it is extremely filled with garbage and single-use plastics (Nayal and Suthar, 2022). The river brings an estimated 1.2 billion pounds of plastic debris into the Bay of Bengal every year (Singh et al., 2025).

There are wide-ranging effects of this pollution. Because they consume plastic particles and experience habitat deterioration, aquatic species upset the equilibrium of the ecosystem (Sarkar et al., 2022). Additionally, people who depend on the Ganges have found microplastics in their drinking water, which poses health problems because plastics release harmful chemicals into the water (Nelms et al., 2021; Singh et al., 2025). The Indian government's National Ganga River Basin Authority is working to address this problem by reducing pollution through better waste management and public awareness initiatives (Gupta et al., 2024). However, issues with enforcement and infrastructure restrict these projects 'efficacy. The Ganges' plastic pollution must be reduced through comprehensive measures that include stronger waste management laws, international collaboration, and technical advancements like recycling programs for plastic garbage and river clean-up projects (Rajan et al., 2023).

Plastics are vitamins for modern businesses, but they pose significant risks to the environment due to their harmful effects on humans, disposal issues, and immortal nature. The influence of these factors on urban environments, air quality, and both marine and terrestrial ecosystems underscores the pressing necessity for sustainable solutions. Considering the challenges faced by our wildlife-disrupted ecosystems and the contributions of plastic waste to greenhouse gas emissions are significant concerns that require collective attention. To effectively address these issues, it is essential to embrace the principles of a circular economy, implement strict regulations on waste management, and invest in the development of biodegradable materials. Through fostering international collaboration, promoting responsible consumption, and enhancing our recycling systems, we have the opportunity to collectively mitigate the environmental impacts of plastics and advance toward a more sustainable future for everyone.

1.2 Plastics, the Environment and Human Health

Many of the items we utilize on a daily basis are composed of plastic, such as packaging, utensils, toys, and various household products. Although plastics offer convenience through their durability, flexibility, and low cost, these same properties also contribute to their potential health hazards. Plastics contain several hazardous substances that are detrimental to human health, such as monomeric building blocks (Bisphenol-A), additives (plasticizers), polyfluorinated chemicals, phthalates, antimony trioxide, and brominated flame retardants (Proshad et al., 2017). Toxic chemicals generated during the plastic production process can have negative health effects when they come into contact with food packaging materials and children's plastic toys. Ingestion, skin contact, inhalation and food are the main sources of exposure to harmful chemicals in humans (Evode et al., 2021). A contaminant known as microplastic can bioaccumulate in the food chain after being consumed by freshwater and marine life, which can cause health problems for humans. Most of the additives present in plastics are carcinogens and endocrine disruptors. Dermatitis is caused by skin contact with some additives that are present in plastics (Araújo et al., 2002).

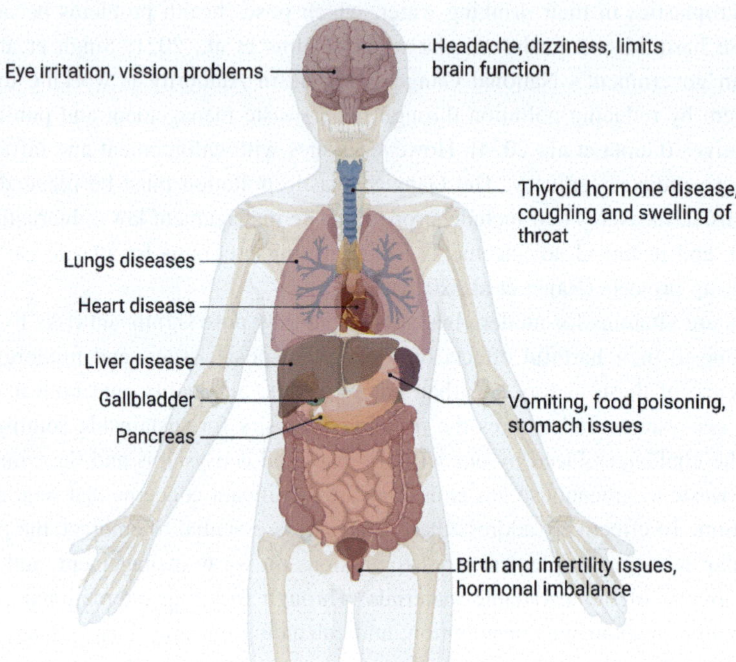

Fig. 1.2 Effect of plastic waste on human health (*Created by using Biorender.com*)

Different additives are used in plastic production, which have various detrimental effects on humans; these impacts are illustrated in Fig. 1.2 and systematically outlined in Table 1.1.

1.2.1 Bisphenol-A (BPA)

Plastics release a range of hazardous chemicals, among which Bisphenol-A (BPA) is particularly significant. BPA serves as a monomeric component in the synthesis of polycarbonate and functions as an additive in polyvinyl chloride (PVC) (Manzoor et al., 2022). As of 2024, global BPA production reached approximately 8.18 million tons annually. Common applications of BPA include baby bottles, the internal coatings of food and beverage cans, and reusable water containers. During the polymerization process, incomplete binding of monomers results in residual BPA, which can subsequently migrate into food and liquids stored in such containers. This leaching process is intensified by repeated washing, as well as the storage of acidic or basic substances at elevated temperatures (Manzoor et al., 2022). The primary routes of human exposure to BPA are ingestion and

Table 1.1 Chemicals and additives in plastic production-associated health effects and plastic types

Sr. no	Toxic chemicals, additives	Plastic types	Uses	Source of exposure	Impact on human health	References
1	Dimethyl phthalate (DMP)	Polyvinyl chloride (PVC), Polystyrene (PS)	Plasticizers	Industrials: In petroleum production, Insect repellents, plastic	Fatal toxicity affects the musculoskeletal system, causing Developmental abnormalities of the eye and ear	Kumar (2018)
2	Di-*n*-butyl phthalate (DBP)	PVC	Adhesive, Industrial solvents	Personal care products, Industrial solvents, Medications	Decreases sperm production, Mild skin irritation	Rajmohan et al., (2019)
3	Bisphenol A	Polyvinyl chloride (PVC), Polycarbonate (PC)	Plasticizers, can liner	Through beverages and packing foods	Mimics oestrogen, Ovarian disorder, obesity, recurrent miscarriages, endometrial hyperplasia	Halden (2010a)
4	Nonylphenol	PVC, cleaners	Anti-static, anti-fog, surfactant (in detergents)	Ingestion of contaminated water and food, especially seafood, and contact with products	Mimics oestrogen	Okunola et al., (2019a)
5	Dioxins	All types of plastics	Release during low-temperature combustion of PVC	Contact and ingestion	Carcinogenic, interferes with testosterone	Okunola et al., (2019a)
6	Polychlorinated biphenyls (PCBs)	All plastics	Dielectrics in electrical equipment	Ingestion of PCBs contaminated marine food	Reproductive disorder alters hormone level	Halden (2010b)

(continued)

Table 1.1 (continued)

Sr. no	Toxic chemicals, additives	Plastic types	Uses	Source of exposure	Impact on human health	References
7	Polycyclic aromatic hydrocarbon	Pesticides, plastics	Manufacturing of pesticides	Inhalation, ingestion, and dermal contact	Developmental and reproductive toxicity	Halden (2010b)
8	Brominated Flame Retardants	Electronic Thermoplastics like computers, cables	Flame retardants	Diet, indoor dust, and inhalation of contaminated air	Possible neurological and reproductive damage	Halden (2010b)
9	Polychlorinated dibenzo-dioxin	PVC	Plasticizer	Contaminated food	Carcinogenic, irritates the skin, eyes and respiratory system. It damages the circulatory, digestive and nervous system, liver, bone marrow	Meeker et al., (2009c)
10	Toluene	Resin, polyurethane, Nylon	Solvent	Inhalation of contaminated air, skin contact	Irritates the eyes and the respiratory tract, which can cause depression	Gilpin et al., (2003)

inhalation. Even trace amounts of BPA migrating from packaging materials into consumables have the potential to interfere with the body's endocrine signaling mechanisms (Proshad et al., 2017). Bisphenol-A (BPA) exposure has been associated with a spectrum of adverse health effects. These include metabolic disorders such as obesity, type 2 diabetes mellitus, and cardiovascular disorders. Apart from this, BPA poses significant reproductive complications to males, such as reduced sperm count with declined production, abnormal development of the penis and urethra, early onset of puberty, and potential links to neurodevelopmental disorders such as autism. In females, it causes endocrine disruption, leading to conditions such as endometrial hyperplasia, polycystic ovarian syndrome (PCOS), infertility, ovarian chromosomal abnormalities, recurrent loss in pregnancy, and breast and prostate cancers. Furthermore, BPA influences the expression of genes involved in the thyroid hormone regulatory axis, thereby altering key metabolic and developmental biological pathways (Mathieu-Denoncourt et al., 2015). Bisphenol-A (BPA) concentrations in human blood and tissues typically range between 0.1 and 10 μg/L, and can be detected in both serum and urine samples. Women of reproductive age and children exhibit higher vulnerability to BPA exposure compared to the general adult population. Consequently, the presence of BPA in the environment poses a significant concern, particularly due to its potential impact on developing fetuses and young children during critical stages of growth and physiological development (Vandenberg et al., 2009; Saal et al., 2007).

1.2.2 Phthalates

Chemically, the phthalates are known to diesters (usually dialkyl or aryl) of 1,2-benzenedicarboxylic acid (phthalic acid), exhibit varying properties depending on the length and branching of alkyl or aryl alcoholic side chains of diesters. Among them, di(2-ethylhexyl) phthalate (DEHP) is produced in large quantities, approximately two million tons annually and is predominantly utilized in the manufacture of medical devices. Beyond medical applications, phthalates serve as solvents in lacquers and coatings and are widely incorporated into personal care products, food packaging, vinyl flooring, toys, and waterproof clothing (Meeker et al., 2009a). Human exposure to phthalates occurs through several pathways, including the use of phthalate-containing medical equipment during procedures such as dialysis, blood transfusion, and extracorporeal membrane oxygenation (ECMO). Additional exposure routes include ingestion of contaminated food, inhalation of household dust, and contact with polluted materials (Meeker et al., 2009b). Due to the absence of covalent bonding between phthalates and the polymer matrix, these compounds readily leach into the surrounding environment. Infants and young children are particularly vulnerable to phthalate exposure through skin contact with contaminated materials and mouthing of plastic toys, as well as ingestion via breastmilk, cow milk, or food packaged in phthalate-laden materials. Elevated phthalate levels are associated with endocrine disruption and congenital abnormalities. In this regard, Duty, Susan M and co-authors reported that phthalate exposure to children's leads to respiratory and dermatological conditions such as rhinitis and eczema (Duty et al., 2005).

1.2.3 Polychlorinated Biphenyls (PCBs)

Polychlorinated biphenyls (PCBs) are POPs that contaminate the marine food web, significantly affecting fish and seabirds. The use of seafood by humans contaminated with PCBs has been linked to an increased risk of disease progression, hormonal imbalances, reproductive dysfunction, and mortality (Ryan et al., 1988). Notably, ingestion of plastic debris by marine organisms, such as the great shearwater (Puffinus gravis), has resulted in significant tissue accumulation of PCBs (Rochman et al., 2013). In humans, PCB exposure has been associated with a broad spectrum of health issues, including cachexia (wasting syndrome), ocular disorders, tumor formation, immune with CNS disruptions, reduced birth weight, reproductive toxicity, and effect on the metabolism of vitamin A.

1.2.4 Halogenated Flame Retardants

Flame retardants (FRs) are added to various products, such as furniture, clothing, and electronics, to improve safety by reducing flammability, suppressing ignition, and slowing fire spread. Key mechanisms include capturing reactive radicals in the flame and forming an insulating char layer that suppress heat and oxygen (Ngo, 2019; Shen et al., 2021). Organic flame retardants are broadly categorized into halogenated and organophosphorus compounds or a combination of both. For instance, *tris*(2,3-dibromopropyl) phosphate contains both halogen and phosphorus elements (Yang et al., 2019). Commonly used halogenated FRs include brominated ethers (e.g., decabromodiphenyl ether), brominated alkanes (e.g., hexabromocyclododecane, HBCD), brominated phenols (e.g., tetrabromobisphenol A, TBBPA), phenyl phosphates (e.g., tricresyl phosphate, TCP), and halogenated organophosphates (e.g., tris(1,3-dichloro-2-propyl) phosphate, TDCPP).

These compounds are environmentally pervasive, entering air, water, and soil systems through degradation, leaching, and emissions during manufacturing, use, and disposal. Accidental releases, such as during fires or industrial accidents, further contribute to their environmental burden (Reemtsma et al., 2006). For example, (Yin et al., 2020) reported median concentrations of polybrominated diphenyl ethers (PBDEs) in Chinese rivers ranging from 0.09 to 2.19 μg/kg (Yin et al., 2020). Inhalation, dermal contact, and inadvertent ingestion, especially in indoor environments, are the predominant exposure pathways for humans (Chen et al., 2020; Li et al., 2023).

1.2.4.1 Widespread Use and Environmental Release

Flame retardants are extensively utilized to meet increasingly stringent fire safety regulations, particularly in polymers that are intrinsically flammable. Global consumption is estimated in millions of tons annually (Ekpe et al., 2020). Consequently, human exposure arises through direct contact with FR-treated products, leaching from aging or discarded materials, occupational settings (e.g., electronics manufacturing, firefighting), and environmental and dietary pathways.

1.2.4.2 Environmental Persistence and Bioaccumulation

Due to their chemical structure—characterized by halogens, phosphorus, or sulfur—FRs exhibit high resistance to oxidative, photolytic, and biological degradation. These properties allow them to persist in the environment and bioaccumulate in living organisms, especially in lipid-rich tissues, placing them within the broader category of persistent organic pollutants (POPs) (Koch et al., 2016; Yang et al., 2022).

1.2.4.3 Structural Similarity to Known Toxicants

Many FRs share structural features with other well-characterized toxic POPs, such as organochlorine pesticides (e.g., mirex, aldrin), polychlorinated biphenyls (PCBs), and organophosphate insecticides (e.g., parathion, chlorpyrifos) (Fishbein, 1974; Lal and

Saxena, 1982; Mali et al., 2023). This similarity has directed toxicological attention toward FRs, leveraging existing models and analytical tools developed for these related compounds.

1.2.4.4 Demonstrated Health Impacts and Regulatory Gaps

Mounting scientific evidence links FR exposure to a range of adverse biological effects, particularly targeting reproductive and immune systems. Although early industry narratives framed FRs as inert, accumulating data from the 2000s onward shifted scientific and regulatory perspectives (Pijnenburg et al., 1995; Santillo & Johnston, 2003). As certain compounds have been phased out, the industry response has involved the rapid introduction of structurally novel FRs, complicating comprehensive toxicological assessment.

FRs act directly on various cellular components, which leads to numerous documented physiological outcomes, especially within the reproductive and immune systems. However, new and more in-depth research is needed on actual biochemical and biophysical modes of action of FR molecules in living systems, including mechanisms of biological membrane disruption, hydrophobic interactions with cellular components, as well as direct and indirect genotoxicity.

1.2.5 Emissions from Plastic Combustion

The incineration or uncontrolled burning of plastic waste releases a range of hazardous air pollutants that pose serious risks to both environmental and human health. These emissions are associated with a variety of acute and chronic health effects, including carcinogenesis, teratogenic outcomes, reproductive system impairments, and respiratory disorders.

Combustion of plastics results in the release of numerous toxic substances, including heavy metals (e.g., mercury, lead, cadmium), persistent organic pollutants such as polychlorinated dibenzo-p-dioxins (PCDDs) and furans, polycyclic aromatic hydrocarbons (PAHs), acid gases (e.g., sulfur dioxide [SO_2], hydrogen chloride [HCl]), particulate matter (dust and grit), NO_x, carbon monoxide (CO), and carbon dioxide (CO_2).

Carbon monoxide, a common by-product of incomplete combustion, can cause symptoms that mimic influenza, including headache, fatigue, dizziness, and nausea. At higher exposure levels, CO may lead to cognitive disorientation, unconsciousness, and in extreme cases, fatal asphyxiation. Dioxins released during plastic combustion are among the most toxic anthropogenic chemicals. They have been linked to a wide array of adverse outcomes, including endocrine disruption, endometriosis, cognitive and motor dysfunctions, ischemic heart disease, and increased risk of hormone-related cancers, particularly breast cancer. The International Agency for Research on Cancer (IARC) classifies dioxins as Group 1 carcinogens, known to be carcinogenic to humans.

Sulfur dioxide exposure primarily affects the upper respiratory tract, causing irritation, coughing, dyspnea, and ocular discomfort. Toluene, another common emission from plastic incineration, can cause reversible toxic effects on the liver, kidneys, and central nervous system, with neurotoxic manifestations often being the most pronounced. Polystyrene-based plastics, when improperly burned or disposed of, release hazardous chemicals that can lead to severe health complications, including emphysema, asthma, cardiovascular dysfunction, hepatotoxicity, nephrotoxicity, vomiting, and reproductive toxicity (Sarkingobir et al. 2021). The cumulative impact of these emissions highlights the need for stringent control measures in plastic waste management systems to mitigate human exposure and reduce ecological harm.

1.3 Plastic Waste Pollution and Sustainable Solutions

Packaging and consumer goods make extensive use of plastic, which serves many valuable industrial purposes. However, this has resulted in a marked increase in plastic waste. The majority of plastic used for these applications comes from non-renewable fossil fuels, significantly contributing to global carbon emissions. As outlined in Sect. 1.1.1, we discuss traditional waste management practices, including landfilling and incineration. While these methods are commonly used, there is an opportunity to explore more sustainable alternatives that can benefit both the environment and future generations.

Recycling and upcycling are two important strategies for achieving sustainability. Recycling involves converting waste into new, usable materials, while upcycling takes this a step further by transforming waste into valuable or functional products. Both methods help to reduce the amount of plastic waste generated and lower the demand for the production of virgin plastic.

Plastic recyclables are generally classified into three main types: primary, secondary, and tertiary recycling. Primary recycling refers to the process of reusing plastic in its original form without altering significant changes to its structure. This can involve melting and remolding used plastic into new products. In contrast, secondary recycling focuses on mechanical processes, such as shredding, granulating, and extruding, that convert plastics into novel materials, although the quality of these materials is typically reduced. Tertiary recycling is the most advanced form that involves the chemical conversion of polymeric plastic into its monomeric building blocks, which allows the preparation of virgin, high-quality polymers. Tertiary recycling is advantageous in terms of having the potential for high-quality outcomes, but it is not universally applicable to all kinds of plastics and is also not economically viable. Amongst the discussed recycling methodologies, primary and secondary recycling is more common and cost-effective but often yields lower-quality materials, which can restrict their use in high-demand applications. Hence, a combination of these methodologies alongside newer technologies and policy support is critical to developing effective and sustainable plastic waste management solutions (Balu et al. 2022).

Despite ongoing challenges in combating plastic waste pollution, a comprehensive approach offers hope. Individuals, businesses, and governments must work together by adopting innovative solutions, reducing plastic use, and encouraging responsible disposal practices. This urgent collective action is essential to protect the environment for future generations.

1.3.1 Reduction and Prevention Strategies

1.3.1.1 Bans on Single-Use Plastics

To address plastic pollution, the government have enacted laws to reduce single-use plastics, such as bags, straws, and cutlery, which harm the environment. While these bans have decreased plastic waste, their success relies on effective enforcement, public awareness, and sustainable alternatives.

In 2019, the EU implemented a directive banning certain single-use plastics with the aim of enforcing this ban by 2021. According to the Environment Agency (EA) reports 2023, this initiative led to a reduction of up to 30% of plastic debris in various coastal areas of the EU (Kiessling et al., 2023). Similarly, Rwanda enacted a nationwide plastic bag ban in 2008, reducing urban plastic waste by approximately 70%, establishing the country as a leader in plastic waste governance (Behuria, 2021; Kasznik and Łapniewska, 2023).

India, generating around 3.5 million metric tons of plastic waste annually, implemented the Plastic Waste Management Amendment Rules (2021), banning certain single-use plastic items (Kumar et al., 2025). Despite this, uneven enforcement and systemic issues, such as limited infrastructure for waste segregation, low public engagement, and the cost-effectiveness of conventional plastics, have hampered widespread impact (Dong et al., 2025; Nøklebye et al., 2023; Saif et al., 2025). For example, Delhi still reports 20–25% of its waste stream comprising banned plastic items (Nøklebye et al., 2023). In the United States, single-use plastic bans are applied at the state and local levels. California, a pioneer in this space, achieved a 72% reduction in plastic bag litter within three years of its 2014 ban (Wagner, 2017).

These cases underscore the effectiveness of well-enforced bans in reducing plastic pollution. However, long-term sustainability requires a multifaceted approach that extends beyond prohibition. Strategies must incorporate innovative recycling technologies, the development of cost-effective biodegradable materials, and comprehensive waste management systems supported by public education and global cooperation (Taylor, 2019).

1.3.1.2 Consumer Awareness and Behaviour Change

Consumer behavior is a major driver of plastic waste, often influenced by convenience, cost, and availability. Raising awareness and encouraging sustainable habits, like using reusable bags, bottles, and containers, can significantly reduce plastic use (Oludoye et al., 2024).

Governments and organizations have launched campaigns to educate the public about the environmental and health risks of plastic. Initiatives like Plastic Free July and India's Swachh Bharat Abhiyan promote reduced plastic consumption and proper waste management (Allison et al., 2022; Luo and Zhao, 2023; Waters et al., 2023).

Eco-labels help guide environmentally conscious choices by highlighting products that are recyclable or biodegradable. In response, businesses are adopting sustainable packaging and increasing transparency to meet rising consumer demand for greener products (Mori et al., 2024; Nøklebye et al., 2023). Incentive-based programs, such as discounts for reusable cups or deposit-return schemes for bottles, have proven effective in changing habits (Lee and Lee, 2024; Mori et al., 2024). Additionally, local initiatives like refill stations, bulk stores, and zero-waste workshops empower communities to reduce plastic reliance (Husin et al., 2025).

Businesses also play a key role by eliminating excess packaging and rewarding sustainable behaviors. Through awareness, incentives, and accessible alternatives, long-term behavior change is possible, fostering a culture of responsible consumption and environmental care (Mori et al., 2024).

1.3.1.3 Sustainable Product Design

As plastic pollution becomes a pressing global concern, industries are shifting toward sustainable product design by adopting eco-friendly materials and innovative manufacturing techniques. Rising consumer demand has accelerated the development of biodegradable, compostable, and recyclable alternatives, though widespread adoption still depends on supportive infrastructure and cross-sector collaboration (Rosenboom et al., 2022).

Key advancements include biodegradable plastics made from natural sources like starch, PLA, and PHA, which decompose through microbial activity. Materials like seaweed-based packaging and agricultural bio-composites offer promising alternatives but require proper disposal in industrial composting facilities to be truly effective (Samir et al., 2022; Barmpaki et al., 2024; Li et al., 2024 Li et al., 2024).

Bio-based and fiber-based packaging made from sugarcane, algae, or recycled paper reduces both plastic use and carbon emissions, supporting circular economy principles. Companies are increasingly turning to molded fiber and plant-based packaging as practical replacements for traditional plastics (Rosenboom et al., 2022). However, challenges remain, particularly around cost and commercial scalability. Many biodegradable materials are more expensive to produce, limiting their availability. Increased R&D is needed to enhance their affordability and performance. Public–private partnerships and regulatory incentives like subsidies and tax breaks can drive broader industry adoption (Arora et al., 2023; Tennakoon et al., 2023).

Crucially, waste management infrastructure must keep pace with material innovation. Many sustainable materials need specific industrial composting conditions to decompose properly; without these, they may end up in landfills or contaminate recycling streams (Souza and Gupta, 2024; Samir et al., 2022).

To fully realize the potential of sustainable product design, cooperation among industries, governments, and consumers is essential. Integrating sustainability into manufacturing processes and encouraging responsible disposal will significantly reduce plastic waste and help pave the way for a cleaner, more eco-conscious future.

1.3.1.4 Corporate Responsibility and Supply Chain Innovations

As plastic pollution grows, industries are increasingly adopting sustainable practices across their supply chains. Many are pledging to make their packaging fully recyclable, reusable, or compostable, aligning with consumer demand and regulatory pressures to reduce environmental harm (Shahzabeen et al., 2023).

A major strategy is the shift toward circular economy models, where businesses collect and repurpose plastic waste into new products. By utilizing post-consumer recycled (PCR) materials, one can reduce the reliance on virgin plastics and lower carbon emissions. Additionally, alternative packaging options, such as biodegradable polymer-based, paper-based, or refillable packaging materials, are gaining popularity. Many retailers are now introducing in-store refill stations and deposit-return programs to promote sustainable consumer habits (Emadian et al., 2017; Omer and Hassan, 2024).

Companies are now publishing sustainability reports that detail their plastic reduction goals and environmental impacts. Policies like Extended Producer Responsibility (EPR) and environmental certifications encourage greener practices (Tumu et al., 2023). These efforts help minimize waste, build trust, improve brand image, and enhance competitiveness. By aligning supply chain innovations with environmental goals, corporations can significantly reduce plastic pollution and drive the transition to a sustainable global economy (Phelan et al., 2022).

1.3.1.5 Economic and Policy Incentives

Economic and policy-based incentives play a vital role in reducing plastic waste by encouraging both businesses and consumers to adopt sustainable practices. Tools like plastic levies, recycling rewards, and subsidies for eco-friendly materials help drive the transition toward a circular economy (Chenavaz and Dimitrov, 2024; Singh and Walker, 2024). By taxing single-use plastics, governments make sustainable alternatives more financially appealing, while grants and funding support innovation in biodegradable packaging and advanced recycling technologies (Issifu et al., 2025).

A standout policy is EPR, which requires manufacturers to manage the full lifecycle of their products, including collection, recycling, and disposal. EPR not only shifts the financial burden of waste management but also encourages companies to design products that are easier to recycle and generate less waste. Countries like Germany and Canada have shown improved recycling rates and waste reduction under strong EPR frameworks (Rubio et al., 2019; Tumu et al., 2023).

Governments must enforce these policies effectively and offer tax incentives or subsidies to companies adopting circular models such as take-back programs and reusable packaging (Gómez-Sanabria and Lindl, 2024; Wang et al., 2020). By integrating financial incentives with strict regulation, policymakers can support green innovation, encourage behavioral change, and help pave the way for a more sustainable, plastic-free economy (Gómez-Sanabria and Lindl, 2024).

1.4 Plastic Waste Management Strategies in Practice

Effectively managing plastic waste requires a comprehensive approach that includes legislation, technology, public engagement, and economic incentives. Despite growing awareness, challenges like poor waste collection systems and limited recycling infrastructure still impede progress, especially in developing countries (Rafey and Siddiqui, 2023).

Improving waste collection and advanced recycling technologies like pyrolysis and chemical recycling offer promise but require investment and regulatory support (Qian et al., 2025). Transitioning from a linear to a circular plastic economy is crucial, involving the development of biodegradable materials, recyclable product design, and EPR policies that hold manufacturers accountable (Andreasi Bassi et al., 2020; Babaremu et al., 2022; Gracida-Alvarez et al., 2023; Tumu et al., 2023). Public participation is another key pillar, with education, incentives, and digital tools enhancing awareness and proper disposal habits (Etim, 2024; Mathew et al., 2023; Ramos, 2024). Economically, the recycling sector needs tax breaks, subsidies, and support through public–private partnerships to compete with virgin plastic production (Milios, 2021; Santos et al., 2025). Emerging technologies such as AI-based sorting systems and blockchain tracking are revolutionizing waste management efficiency and transparency (Garcia and Robertson, 2017; Geyer et al., 2017). Sustainable product design, corporate-led initiatives, and grassroots efforts by NGOs and communities are also driving meaningful change (Payne and Jones, 2021; N. S., 2022). Together, these multi-layered strategies can create a resilient and sustainable plastic waste management ecosystem (Andía et al., 2022; Barford and Ahmad, 2021). Figure 1.3 provides the data on annual plastic waste generated and its disposal by disposal method.

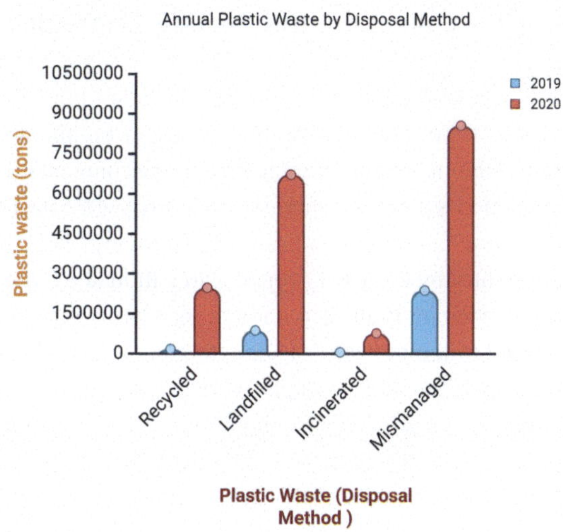

Fig. 1.3 Annual plastic waste by disposal method (2000–2019) (*Created by using Biorender.com*)

1.4.1 Reduction and Prevention Strategies

Reducing plastic production and consumption is a foundational strategy in managing plastic waste. Policy measures like bans on single-use plastics have been instrumental in this effort. For instance, India's nationwide ban and the EU's Single-Use Plastics Directive aim to cut down on disposable plastic items and promote EPR programs (Simantiris, 2024; Tan et al., 2021). Alternatives to conventional plastics, such as biodegradable and bio-based materials, are also gaining traction. Innovations include seaweed-based packaging by Notpla in the UK and sorghum-based edible cutlery by Indian company Bakeys, reducing dependence on petroleum-based plastics (Knoblauch and Mederake, 2021; Tan et al., 2021).

Educational campaigns, incentives for reusable products, and community actions, like Kenya's cloth bag drive and Starbucks' discounts for reusable cups, demonstrate effective strategies for public engagement (Herberz et al.,, 2020; Tan et al., 2021; Moshood et al., 2022). Global movements such as Plastic-Free July and local actions like beach clean-ups also contribute to shifting societal norms (Marathe and Bank, 2022; Rabiu and Jaeger-Erben, 2024). The Alliance to End Plastic Waste, comprising companies like Procter & Gamble and PepsiCo, supports waste reduction solutions worldwide (Bank et al., 2021; Filimonova et al., 2023). Additionally, brands like Adidas and Patagonia showcase how recycled plastics can be converted into sustainable consumer products (Filimonova et al., 2023; Lampitt et al., 2023).

Ultimately, achieving a sustainable plastic economy depends not only on active collaboration among governments, industries, and individuals but also, through bans, innovation, and awareness, can minimize environmental pollution and promote responsible resource use (Filimonova et al., 2023).

1.4.2 Collection and Sorting Strategies

Efficient collection and sorting systems are fundamental to sustainable plastic waste management. Structured collection strategies significantly improve the volume and quality of recyclable materials, thus reducing environmental pollution (Bank et al., 2021). Curbside collection schemes, where households segregate recyclables for municipal pickup, have proven effective in countries like Germany. The Green Dot system has led to high recycling rates by ensuring proper segregation at the source (Vukicevic et al., 2025). Similarly, cities like San Francisco enforce mandatory recycling laws to reduce landfill dependence.

In developing nations, strengthening informal waste collection is essential. Waste pickers play a crucial role in plastic recovery but often work in unsafe and informal conditions (Beghetto et al., 2023; Aydın et al., 2025). Integrating them into formal systems through cooperatives like Brazil's COOPAMARE or partnerships like India's Solid Waste Collection and Handling (SWaCH) improves recycling outcomes and worker livelihoods (Beswick-Parsons et al., 2023). Technological innovation is enhancing waste collection and sorting. Smart bins with sensors and AI systems, like those in the Netherlands, efficiently identify and classify plastics (Lange, 2021).

The effectiveness of collection and sorting strategies relies on forward-thinking regulations, technology, and public participation. A hybrid approach combining formal infrastructure and informal contributions offers the most sustainable path forward (Vukicevic et al., 2025).

1.4.3 Recycling and Reuse Strategies

Recycling and reuse are central to reducing plastic waste and advancing a circular economy. As discussed in Sect. 1.3, the key approaches of recycling include mechanical, chemical recycling, and upcycling/downcycling.

Mechanical recycling involves shredding, cleaning, and remelting plastics to produce new products. It is widely applied to materials like polypropylene (PP), polyethylene terephthalate (PET), and polyethylene (PE) (Aydonat et al., 2024). In India, companies such as Reliance Industries and Ganesha Ecosphere operate major PET recycling plants, supplying recycled polyester for textiles. However, contamination and quality degradation over multiple cycles limit its effectiveness (Lange, 2021). Chemical recycling breaks down polymeric plastics into monomers or basic building functionality, enabling the synthesis of new materials (Coates and Getzler, 2020). Technologies like pyrolysis and depolymerization convert plastic waste into virgin-equivalent resins. Indian companies such as Banyan Nation and Gravita India are making significant investments in chemical recycling to transform plastic waste into reusable raw materials (Lange, 2021; Vollmer et al., 2020).

Upcycling turns waste plastics into valuable products like fashion accessories and building materials. Aarohana EcoSocial in India, for instance, creates textiles from recycled plastic waste (Coates and Getzler, 2020). In contrast, downcycling transforms plastics into lower-value products like insulation or plastic lumber.

1.4.4 Waste-To-Energy Technologies

Waste-to-Energy (WTE) technologies provide an innovative solution for managing non-recyclable plastics by converting them into usable energy, thereby reducing reliance on landfills. Two main WTE methods include incineration with energy recovery and thermochemical processes such as gasification and pyrolysis.

Incineration with Energy Recovery burns plastic waste at high temperatures to generate heat and electricity (Li et al., 2024). Advanced incineration facilities in countries like Sweden and Japan incorporate pollution control systems to minimize environmental impact, ensuring emissions like dioxins and furans stay within legal limits (Rahman et al., 2025). In India, the Okhla and Ghazipur WTE plants in Delhi process thousands of tons of municipal waste daily, producing 16 MW and 12 MW of electricity, respectively (Pannerselvam et al., 2022). As outlined earlier, air pollution remains a concern; technologies such as carbon capture, electrostatic precipitators, and flue gas treatment have significantly improved emissions control (Karmakar et al., 2023).

Gasification and pyrolysis convert plastic waste under oxygen-limited conditions into fuels and chemical feedstocks (Qureshi et al., 2020). Gasification produces syngas, a mix of hydrogen and carbon monoxide which is further used in energy generation or chemical manufacturing (Saebea et al., 2020; Bashir et al., 2025). The CSIR-Indian Institute of Petroleum in Dehradun has pioneered methods to convert mixed plastic waste into valuable fuels (Lopez et al., 2018). Pyrolysis, which heats plastic between 300–600 °C without oxygen, produces pyrolysis oil, synthetic gas, and char (Saad and Williams, 2016; Qureshi et al., 2020). Recent advancements in pyrolysis and gasification are improving energy recovery efficiency and environmental sustainability. These technologies are especially crucial in India's context, where the sheer volume of non-recyclable plastic waste demands scalable and innovative thermal conversion methods.

Microwave-assisted pyrolysis (MAP) is gaining attention for replacing traditional external heat sources with microwave heating, resulting in better energy efficiency, uniform heat distribution, and higher-quality fuel outputs (Suriapparao et al., 2022). Unlike conventional pyrolysis, MAP reduces partial decomposition by ensuring even heating. Indian institutions like IIT Delhi and CSIR-IIP have shown that MAP can efficiently convert mixed plastic waste into high-value fuels, presenting a cost-effective alternative to petroleum-based fuels (Suriapparao et al., 2022). This is particularly impactful in metropolitan regions like Delhi and Mumbai, where landfill accumulation of non-recyclable plastics is severe (Attallah et al., 2024).

1.5 Conclusion

Plastic waste has become a major environmental and public health issue due to its widespread use and resistance to degradation. It affects land, water, and air ecosystems while threatening human health from chemical exposure and microplastics in food and water. Traditional disposal methods like landfilling and incineration often worsen the problem by releasing hazardous emissions.

This chapter outlines sustainable strategies for managing plastic waste, including source reduction, changes in consumer behavior, and circular economy models. Effective waste collection, sorting, and recycling methods are vital, and innovative approaches like upcycling and pyrolysis offer solutions for converting waste into energy and valuable materials.

A collaborative response is essential. Governments must enforce regulations, industries should adopt sustainable practices, and consumers need education and incentives for responsible consumption. Public–private partnerships and international cooperation will be key in scaling solutions and investing in research for biodegradable materials.

Transitioning to a circular economy can minimize environmental risks while fostering innovation and creating green jobs. Though the challenges are significant, coordinated action can lead to a sustainable future.

References

Allison, A. L., Baird, H. M., Lorencatto, F., Webb, T. L. & Michie, S. (2022). Reducing plastic waste: A meta-analysis of influences on behaviour and interventions. *Journal of Cleaner Production, 380*, 134860. https://doi.org/10.1016/J.JCLEPRO.2022.134860.

Andía, D., Charca, S., Reynolds-Cuéllar, P. & Noel, J. (2022). Community-oriented engineering co-design: case studies from the Peruvian Highlands. *Humanities and Social Sciences Communications, 9*(1), 1–9. https://doi.org/10.1057/S41599-022-01331-0.

Andreasi Bassi, S., Boldrin, A., Faraca, G. & Astrup, T. F. (2020). Extended producer responsibility: How to unlock the environmental and economic potential of plastic packaging waste? *Resources, Conservation and Recycling, 162*, 105030. https://doi.org/10.1016/J.RESCONREC.2020.105030.

Araújo, P. H. H., Sayer, C., Giudici, R. & Poço, J. G. R. (2002). Techniques for reducing residual monomer content in polymers: A review. *Polymer Engineering & Science, 42*(7), 1442–1468. https://doi.org/10.1002/pen.11043.

Arora, Y., Sharma, S. & Sharma, V. (2023). Microalgae in bioplastic production: A comprehensive review. *Arabian Journal for Science and Engineering, 48*(6), 7225–7241. https://doi.org/10.1007/S13369-023-07871-0/METRICS.

Attallah, O. A., Ferrero, P., Ljesevic, M., Loncarevic, B., Aleksic, I., Pantelic, B., Gojgic, G., Siaperas, R., Topakas, E., Beskoski, V. & Nikodinovic-Runic, J. (2024). Micro-wave induced pyrolysis of low density polyethylene (LDPE) and biodegradation of resulting wax in soil and by defined microbial consortia is closing the loop towards LDPE upcycling. *Journal of Environmental Chemical Engineering, 12*(6), 114269. https://doi.org/10.1016/J.JECE.2024.114269.

Aydın, S., Güneysu, S., Ciner, M. N., Özbaş, E. E., Ozcan, H. K. & Öngen, A. (2025). The effect of volume reduction methods on beverage packaging waste recycling in the deposit return system. *International Journal of Environmental Science and Technology*, 1–20. https://doi.org/10.1007/S13762-025-06476-4/TABLES/10.

Aydonat, S., Hergesell, A. H., Seitzinger, C. L., Lennarz, R., Chang, G., Sievers, C., Meisner, J., Vollmer, I. & Göstl, R. (2024). Leveraging mechanochemistry for sustainable polymer degradation. *Polymer Journal, 56*(4), 249–268. https://doi.org/10.1038/s41428-023-00863-9.

Babaremu, K. O., Okoya, S. A., Hughes, E., Tijani, B., Teidi, D., Akpan, A., Igwe, J., Karera, S., Oyinlola, M. & Akinlabi, E. T. (2022). Sustainable plastic waste management in a circular economy. *Heliyon, 8*(7), e09984. https://doi.org/10.1016/J.HELIYON.2022.E09984.

Balu, R., Dutta, N. K., & Roy Choudhury, N. (2022). Plastic waste upcycling: A sustainable solution for waste management, product development, and circular economy. *Polymers, 14*(22), 4788. https://doi.org/10.3390/polym14224788.

Bank, M. S., Ok, Y. S., Swarzenski, P. W., Duarte, C. M., Rillig, M. C., Koelmans, A. A., Metian, M., Wright, S., Provencher, J. F., Sanden, M., Jordaan, A., Wagner, M. & Thiel, M. (2021). Global plastic pollution observation system to aid policy. *Environmental Science and Technology, 55*(12), 7770–7775. https://doi.org/10.1021/ACS.EST.1C00818/ASSET/IMAGES/LARGE/ES1C00818_0004.JPEG.

Barboza, L. G. A. & Gimenez, B. C. G. (2015) Microplastics in the marine environment: Current trends and future perspectives. *Marine Pollution Bulletin, 97*(1–2), 5–12. https://doi.org/10.1016/j.marpolbul.2015.06.008.

Barford, A. & Ahmad, S. R. (2021). A call for a socially restorative circular economy: Waste pickers in the recycled plastics supply chain. *Circular Economy and Sustainability, 1*(2), 761–782. https://doi.org/10.1007/S43615-021-00056-7/TABLES/3.

Barmpaki, A. A., Paul, U. C., Nardi, M. & Athanassiou, A. (2024). Eco-friendly blends of polylactic acid and polyhydroxybutyrate enhanced with epoxidized soybean oil methyl ester for food-packaging applications. *ACS Applied Polymer Materials, 6*(15), 8997–9007. https://doi.org/10.1021/ACSAPM.4C01341/ASSET/IMAGES/LARGE/AP4C01341_0005.JPEG.

Bashir, M. A., Ji, T., Weidman, J., Soong, Y., Gray, M., Shi, F. & Wang, P. (2025). Plastic waste gasification for low-carbon hydrogen production: a comprehensive review. *Energy Advances, 4*(3), 330–363. https://doi.org/10.1039/D4YA00292J.

Bastian, L., Yano, J., Hirai, Y. & Sakai, ichi, S. (2013). Behavior of PCDD/Fs during open burning of municipal solid waste in open dumping sites. *Journal of Material Cycles and Waste Management, 15*(2), 229–241. https://doi.org/10.1007/S10163-012-0114-5/FIGURES/6.

Beghetto, V., Gatto, V., Samiolo, R., Scolaro, C., Brahimi, S., Facchin, M. & Visco, A. (2023). Plastics today: Key challenges and EU strategies towards carbon neutrality: A review. *Environmental Pollution, 334*, 122102. https://doi.org/10.1016/J.ENVPOL.2023.122102.

Behuria, P. (2021). Ban the (plastic) bag? Explaining variation in the implementation of plastic bag bans in Rwanda, Kenya and Uganda. *Environment and Planning C: Politics and Space, 39*(8), 1791–1808. https://doi.org/10.1177/2399654421994836.

Beswick-Parsons, R., Jackson, P. & Evans, D. M. (2023). Understanding national variations in reusable packaging: Commercial drivers, regulatory factors, and provisioning systems. *Geoforum, 145*, 103844. https://doi.org/10.1016/J.GEOFORUM.2023.103844.

Carriera, F., Di Fiore, C. & Avino, P. (2024). Occurrence of microplastics in the atmosphere: An overview on sources, analytical challenges, and human health effects. *Atmosphere,* 15(7), 863. https://doi.org/10.3390/ATMOS15070863.

Chen, Y., Liu, Q., Ma, J., Yang, S., Wu, Y. & An, Y. (2020). A review on organophosphate flame retardants in indoor dust from China: Implications for human exposure. *Chemosphere, 260*, 127633. https://doi.org/10.1016/J.CHEMOSPHERE.2020.127633.

Chenavaz, R. Y. & Dimitrov, S. (2024). From waste to wealth: Policies to promote the circular economy. *Journal of Cleaner Production, 443*, 141086. https://doi.org/10.1016/J.JCLEPRO.2024.141086.

Coates, G. W. & Getzler, Y. D. Y. L. (2020). Chemical recycling to monomer for an ideal, circular polymer economy. *Nature Reviews Materials, 5*(7), 501–516. https://doi.org/10.1038/S41578-020-0190-4;SUBJMETA=1028,301,455,638,639,923,959;KWRD=POLYMERIZATION+MECHANISMS,POLYMERS.

Dong, H., Zhang, R., Wang, X., Zeng, J., Chai, L., Niu, X., Xu, L., Zhou, Y., Gong, P. & Yin, Q. (2025). Geographical features and management strategies for microplastic loads in freshwater lakes. *npj Clean Water, 8*(1), 1–11. https://doi.org/10.1038/s41545-025-00459-1.

Duty, S. M., Ackerman, R. M., Calafat, A. M. & Hauser, R. (2005). Personal care product use predicts urinary concentrations of some phthalate monoesters. *Environmental Health Perspectives, 113*(11), 1530–1535. https://doi.org/10.1289/ehp.8083.

Ekpe, O. D., Choo, G., Barceló, D. & Oh, J. E. (2020). Introduction of emerging halogenated flame retardants in the environment. *Comprehensive Analytical Chemistry, 88*, 1–39. https://doi.org/10.1016/BS.COAC.2019.11.002.

Emadian, S. M., Onay, T. T. & Demirel, B. (2017). Biodegradation of bioplastics in natural environments. *Waste Management, 59*, 526–536. https://doi.org/10.1016/j.wasman.2016.10.006.

Eriksen, M., Lebreton, L. C. M., Carson, H. S., Thiel, M., Moore, C. J., Borerro, J. C., Galgani, F., Ryan, P. G. & Reisser, J. (2014). Plastic pollution in the world's oceans: more than 5 trillion plastic pieces weighing over 250,000 tons afloat at sea. *PLoS ONE, 9*(12), e111913. https://doi.org/10.1371/journal.pone.0111913.

Etim, E. (2024) Leveraging public awareness and behavioural change for entrepreneurial waste management. *Heliyon, 10*(21). https://doi.org/10.1016/j.heliyon.2024.e40063.

Evode, N., Qamar, S. A., Bilal, M., Barceló, D. & Iqbal, H. M. N. (2021). Plastic waste and its management strategies for environmental sustainability. *Case Studies in Chemical and Environmental Engineering, 4*, 100142. https://doi.org/10.1016/j.cscee.2021.100142.

Filimonova, I. V., Krivosheeva, O. I. & Mishenin, M. V. (2023). The economic effect of public–private partnerships in the implementation of climate projects for the disposal of municipal solid waste. *Energy Reports, 9*, 996–1002. https://doi.org/10.1016/J.EGYR.2022.11.042.

Fishbein, L. (1974). Mutagens and potential mutagens in the biosphere I. DDT and its metabolites, polychlorinated biphenyls, chlorodioxins, polycyclic aromatic hydrocarbons, haloethers. *Science of The Total Environment, 2*(4), 305–340. https://doi.org/10.1016/0048-9697(74)90001-1.

Garcia, J. M. & Robertson, M. L. (2017). The future of plastics recycling. *Science, 358*(6365), 870–872. https://doi.org/10.1126/SCIENCE.AAQ0324/ASSET/53AC3A00-0BBA-41D7-A05A-BE986DAF5476/ASSETS/GRAPHIC/358_870_F1.JPEG.

Geyer, R., Jambeck, J. R. & Law, K. L. (2017). Production, use, and fate of all plastics ever made. *Science Advances, 3*(7). https://doi.org/10.1126/SCIADV.1700782/SUPPL_FILE/1700782_SM.PDF.

Gilpin, R. K., Wagel, D. J. & Solch, J. G. (2003). Production, distribution, and fate of polychlorinated dibenzo- *p* -dioxins, dibenzofurans, and related organohalogens in the environment. In *Dioxins and Health.* Wiley, pp. 55–87. https://doi.org/10.1002/0471722014.ch2.

Gómez-Sanabria, A. & Lindl, F. (2024). The crucial role of circular waste management systems in cutting waste leakage into aquatic environments. *Nature Communications, 15*(1), 1–13. https://doi.org/10.1038/S41467-024-49555-9;SUBJMETA=172,4081,685,704,844;KWRD=ENVIRONMENTAL+IMPACT,SUSTAINABILITY.

Gracida-Alvarez, U. R., Xu, H., Benavides, P. T., Wang, M. & Hawkins, T. R. (2023). Circular economy sustainability analysis framework for plastics: Application for poly(ethylene Terephthalate)

(PET). *ACS Sustainable Chemistry and Engineering, 11*(2), 514–524. https://doi.org/10.1021/ACSSUSCHEMENG.2C04626/ASSET/IMAGES/LARGE/SC2C04626_0006.JPEG.

Gregory, M. R. (2009). Environmental implications of plastic debris in marine settings—entanglement, ingestion, smothering, hangers-on, hitch-hiking and alien invasions. *Philosophical Transactions of the Royal Society B: Biological Sciences, 364*(1526), 2013–2025. https://doi.org/10.1098/rstb.2008.0265.

Gupta, P., Saha, M., Naik, A., Kumar, M. M., Rathore, C., Vashishth, S., Maitra, S. P., Bhardwaj, K. D. & Thukral, H. (2024). A comprehensive assessment of macro and microplastics from Rivers Ganga and Yamuna: Unveiling the seasonal, spatial and risk factors. *Journal of Hazardous Materials, 469*, 133926. https://doi.org/10.1016/J.JHAZMAT.2024.133926.

Hakeem, I. G., Aberuagba, F. & Musa, U. (2018). Catalytic pyrolysis of waste polypropylene using Ahoko kaolin from Nigeria. *Applied Petrochemical Research, 8*(4), 203–210. https://doi.org/10.1007/s13203-018-0207-8.

Halden, R. U. (2010a). Plastics and health risks. *Annual Review of Public Health, 31*(1), 179–194. https://doi.org/10.1146/annurev.publhealth.012809.103714.

Halden, R. U. (2010b). Plastics and health risks. *Annual Review of Public Health, 31*(1), 179–194. https://doi.org/10.1146/annurev.publhealth.012809.103714.

Herberz, T., Barlow, C. Y. & Finkbeiner, M. (2020). Sustainability assessment of a single-use plastics ban. *Sustainability, 12*(9), 3746. https://doi.org/10.3390/SU12093746.

Husin, A., Helmi, H., Nengsih, Y. K. & Rendana, M. (2025). Environmental education in schools: Sustainability and hope. *Discover Sustainability, 6*(1), 1–11. https://doi.org/10.1007/S43621-025-00837-2/FIGURES/3.

Islam, Md. R., Ruponti, S. A., Rakib, Md. A., Nguyen, H. Q. & Mourshed, M. (2023). Current scenario and challenges of plastic pollution in Bangladesh: a focus on farmlands and terrestrial ecosystems. *Frontiers of Environmental Science & Engineering, 17*(6), 66. https://doi.org/10.1007/s11783-023-1666-4.

Issifu, I., Dahmouni, I. & Sumaila, U. R. (2025). Assessing the ecological and economic transformation pathways of plastic production system. *Journal of Environmental Management, 374*, 124104. https://doi.org/10.1016/J.JENVMAN.2025.124104.

Jansen, M. A. K., Andrady, A. L., Bornman, J. F., Aucamp, P. J., Bais, A. F., Banaszak, A. T., Barnes, P. W., Bernhard, G. H., Bruckman, L. S., Busquets, R., Häder, D. P., Hanson, M. L., Heikkilä, A. M., Hylander, S., Lucas, R. M., Mackenzie, R., Madronich, S., Neale, P. J., Neale, R. E., Olsen, C. M., Ossola, R., Pandey, K. K., Petropavlovskikh, I., Revell, L. E., Robinson, S. A., Robson, T. M., Rose, K. C., Solomon, K. R., Andersen, M. P. S., Sulzberger, B., Wallington, T. J., Wang, Q. W., Wängberg, S. Å., White, C. C., Young, A. R., Zepp, R. G. & Zhu, L. (2024). Plastics in the environment in the context of UV radiation, climate change and the montreal protocol: UNEP environmental effects assessment panel, update 2023. *Photochemical & Photobiological Sciences, 23*(4), 629–650. https://doi.org/10.1007/S43630-024-00552-3.

Karmakar, A., Daftari, T., Sivagami, K., Chandan, M. R., Shaik, A. H., Kiran, B. & Chakraborty, S. (2023). A comprehensive insight into waste to energy conversion strategies in India and its associated air pollution hazard. *Environmental Technology & Innovation, 29*, 103017. https://doi.org/10.1016/J.ETI.2023.103017.

Kasznik, D. & Łapniewska, Z. (2023). The end of plastic? The EU's directive on single-use plastics and its implementation in Poland. *Environmental Science & Policy, 145*, 151–163. https://doi.org/10.1016/J.ENVSCI.2023.04.005.

Kehinde, O., Ramonu, O. J., Babaremu, K. O. & Justin, L. D. (2020). Plastic wastes: Environmental hazard and instrument for wealth creation in Nigeria. *Heliyon, 6*(10), e05131. https://doi.org/10.1016/j.heliyon.2020.e05131.

Kiessling, T., Hinzmann, M., Mederake, L., Dittmann, S., Brennecke, D., Böhm-Beck, M., Knickmeier, K. & Thiel, M. (2023). What potential does the EU single-use plastics directive have for reducing plastic pollution at coastlines and riversides? An evaluation based on citizen science data. *Waste Management*, *164*, 106–118. https://doi.org/10.1016/J.WASMAN.2023.03.042.

Knoblauch, D. & Mederake, L. (2021). Government policies combatting plastic pollution. *Current Opinion in Toxicology*, *28*, 87–96. https://doi.org/10.1016/J.COTOX.2021.10.003.

Koch, C., Dundua, A., Aragon-Gomez, J., Nachev, M., Stephan, S., Willach, S., Ulbricht, M., Schmitz, O. J., Schmidt, T. C. & Sures, B. (2016). Degradation of polymeric brominated flame retardants: Development of an analytical approach using PolyFR and UV irradiation. *Environmental Science and Technology*, *50*(23), 12912–12920. https://doi.org/10.1021/ACS.EST.6B04083/ASSET/IMAGES/LARGE/ES-2016-04083W_0004.JPEG.

Kumar, P. (2018). Role of plastics on human health. *The Indian Journal of Pediatrics*, *85*(5), 384–389. https://doi.org/10.1007/s12098-017-2595-7.

Kumar, S., Sinha, S., Ranjan, R. & Saurav, S. (2025) Optimization reverse logistics for waste plastic recycling in India: A genetic algorithm approach. *Innovative Infrastructure Solutions*, *10*(5), 1–13. https://doi.org/10.1007/S41062-025-01969-0.

Lal, R. & Saxena, D. M. (1982). Accumulation, metabolism, and effects of organochlorine insecticides on microorganisms. *Microbiological Reviews*, *46*(1), 95–127. https://doi.org/10.1128/MR.46.1.95-127.1982/ASSET/E04A58A0-2A5D-43B2-8591-5472DFA58DA1/ASSETS/MR.46.1.95-127.1982.FP.PNG.

Lampitt, R. S., Fletcher, S., Cole, M., Kloker, A., Krause, S., O'Hara, F., Ryde, P., Saha, M., Voronkova, A. & Whyle, A. (2023). Stakeholder alliances are essential to reduce the scourge of plastic pollution. *Nature Communications*, *14*(1), 1–3. https://doi.org/10.1038/S41467-023-38613-3;SUBJMETA=172,2788,4081,499,689,692,704,706;KWRD=DECISION+MAKING,ENVIRONMENTAL+IMPACT,RISK+FACTORS.

Lange, J. P. (2021). Managing plastic waste-sorting, recycling, disposal, and product redesign. *ACS Sustainable Chemistry and Engineering*, *9*(47), 15722–15738. https://doi.org/10.1021/ACSSUSCHEMENG.1C05013/ASSET/IMAGES/LARGE/SC1C05013_0001.JPEG.

Lee, S. E. & Lee, K. H. (2024). Environmentally sustainable fashion and conspicuous behavior. *Humanities and Social Sciences Communications*, *11*(1), 1–10. https://doi.org/10.1057/s41599-024-02955-0.

Li, M., Fei, J., Zhang, Z., Sun, Q. & Liu, C. (2023). Organophosphate esters in Chinese rice: Occurrence, distribution, and human exposure risks. *Science of The Total Environment*, *862*, 160915. https://doi.org/10.1016/J.SCITOTENV.2022.160915.

Li, Z., Fan, T. W., Lun, M. S. & Li, Q. (2024). Optimization of municipal solid waste incineration for low-NOx emissions through numerical simulation. *Scientific Reports*, *14*(1), 1–12. https://doi.org/10.1038/S41598-024-69019-W;SUBJMETA=166,172,638,639,704,898;KWRD=CHEMICAL+ENGINEERING,CHEMISTRY,ENVIRONMENTAL+SCIENCES.

Li, Y. C., Wang, S., Qian, S., Liu, Z., Weng, Y. & Zhang, Y. (2024). Depolymerization and Re/Upcycling of biodegradable PLA plastics. *ACS Omega*, *9*(12), 13509–13521. https://doi.org/10.1021/ACSOMEGA.3C08674/ASSET/IMAGES/LARGE/AO3C08674_0009.JPEG.

Lopez, G., Artetxe, M., Amutio, M., Alvarez, J., Bilbao, J. & Olazar, M. (2018). Recent advances in the gasification of waste plastics. A critical overview. *Renewable and Sustainable Energy Reviews*, *82*, 576–596. https://doi.org/10.1016/J.RSER.2017.09.032.

Luo, Y. & Zhao, J. (2023). Using behavioral interventions to reduce single-use produce bags. *Resources, Conservation and Recycling*, *193*, 106942. https://doi.org/10.1016/J.RESCONREC.2023.106942.

Mali, H., Shah, C., Raghunandan, B. H., Prajapati, A. S., Patel, D. H., Trivedi, U. & Subramanian, R. B. (2023). Organophosphate pesticides an emerging environmental contaminant: Pollution, toxicity, bioremediation progress, and remaining challenges. *Journal of Environmental Sciences, 127*, 234–250. https://doi.org/10.1016/J.JES.2022.04.023.

Manzoor, M. F., Tariq, T., Fatima, B., Sahar, A., Tariq, F., Munir, S., Khan, S., Nawaz Ranjha, M. M. A., Sameen, A., Zeng, X.-A. & Ibrahim, S. A. (2022). An insight into bisphenol A, food exposure and its adverse effects on health: A review. *Frontiers in Nutrition, 9*. https://doi.org/10.3389/fnut.2022.1047827.

Marathe, N. P., Bank, M. S. (2022). The microplastic-antibiotic resistance connection, pp. 311–322. https://doi.org/10.1007/978-3-030-78627-4_9.

Matavos-Aramyan, S. (2024). Addressing the microplastic crisis: A multifaceted approach to removal and regulation. *Environmental Advances, 17*, 100579. https://doi.org/10.1016/j.envadv.2024.100579.

Mathew, A., Isbanner, S., Xi, Y., Rundle-Thiele, S., David, P., Li, G. & Lee, D. (2023). A systematic literature review of voluntary behaviour change approaches in single use plastic reduction. *Journal of Environmental Management, 336*. https://doi.org/10.1016/j.jenvman.2023.117582.

Mathieu-Denoncourt, J., Wallace, S. J., de Solla, S. R. & Langlois, V. S. (2015). Plasticizer endocrine disruption: Highlighting developmental and reproductive effects in mammals and non-mammalian aquatic species. *General and Comparative Endocrinology, 219*, 74–88. https://doi.org/10.1016/j.ygcen.2014.11.003.

Meeker, J. D., Sathyanarayana, S. & Swan, S. H. (2009a). Phthalates and other additives in plastics: human exposure and associated health outcomes. *Philosophical Transactions of the Royal Society B: Biological Sciences, 364*(1526), 2097–2113. https://doi.org/10.1098/rstb.2008.0268.

Meeker, J. D., Sathyanarayana, S. & Swan, S. H. (2009b). Phthalates and other additives in plastics: Human exposure and associated health outcomes. *Philosophical Transactions of the Royal Society B: Biological Sciences, 364*(1526), 2097–2113. https://doi.org/10.1098/rstb.2008.0268.

Meeker, J. D., Sathyanarayana, S. & Swan, S. H. (2009c). Phthalates and other additives in plastics: human exposure and associated health outcomes. *Philosophical Transactions of the Royal Society B: Biological Sciences, 364*(1526), 2097–2113. https://doi.org/10.1098/rstb.2008.0268.

Milios, L. (2021). Towards a circular economy taxation framework: Expectations and challenges of implementation. *Circular Economy and Sustainability, 1*(2), 477–498. https://doi.org/10.1007/S43615-020-00002-Z/FIGURES/2.

Mori, Y., Nakamata, T., Kuwayama, R., Yuki, S. & Ohnuma, S. (2024). Developing the littering behavior model focusing on implementation intention: A challenge to anti-environmental behavior. *Journal of Material Cycles and Waste Management, 26*(2), 776–791. https://doi.org/10.1007/S10163-024-01909-7/FIGURES/5.

Moshood, T. D., Nawanir, G., Mahmud, F., Mohamad, F., Ahmad, M. H. & AbdulGhani, A. (2022). Sustainability of biodegradable plastics: New problem or solution to solve the global plastic pollution? *Current Research in Green and Sustainable Chemistry, 5*, 100273. https://doi.org/10.1016/J.CRGSC.2022.100273.

N. S. (2022). Plastic waste management: A road map to achieve circular economy and recent innovations in pyrolysis. *Science of The Total Environment, 809*, 151160. https://doi.org/10.1016/J.SCITOTENV.2021.151160.

Nayal, R. & Suthar, S. (2022). First report on microplastics in tributaries of the upper Ganga River along Dehradun, India: Quantitative estimation and characterizations. *Journal of Hazardous Materials Advances, 8*, 100190. https://doi.org/10.1016/J.HAZADV.2022.100190.

Nelms, S. E., Duncan, E. M., Patel, S., Badola, R., Bhola, S., Chakma, S., Chowdhury, G. W., Godley, B. J., Haque, A. B., Johnson, J. A., Khatoon, H., Kumar, S., Napper, I. E., Niloy, M. N. H., Akter, T., Badola, S., Dev, A., Rawat, S., Santillo, D., Sarker, S., Sharma, E. & Koldewey,

H. (2021). Riverine plastic pollution from fisheries: Insights from the Ganges River system. *The Science of the total environment, 756*. https://doi.org/10.1016/J.SCITOTENV.2020.143305.

Ngo, T. (2019). Development of sustainable flame-retardant materials. *Green Materials, 8*(3), 101–122. https://doi.org/10.1680/JGRMA.19.00060/ASSET/IMAGES/SMALL/JGRMA8-0101-F15.GIF.

Nøklebye, E., Adam, H. N., Roy-Basu, A., Bharat, G. K. & Steindal, E. H. (2023). Plastic bans in India—Addressing the socio-economic and environmental complexities. *Environmental Science & Policy, 139*, 219–227. https://doi.org/10.1016/J.ENVSCI.2022.11.005.

Okunola, A. A., Kehinde, I, O., Oluwaseun, A. & Olufiropo, E, A. (2019a). Public and environmental health effects of plastic wastes disposal: A review. *Journal of Toxicology and Risk Assessment, 5*(2). https://doi.org/10.23937/2572-4061.1510021.

Oludoye, O. O., Supakata, N., Srithongouthai, S., Kanokkantapong, V., Van den Broucke, S., Ogunyebi, L. & Lubell, M. (2024). Pro-environmental behavior regarding single-use plastics reduction in urban–rural communities of Thailand: Implication for public policy. *Scientific Reports, 14*(1), 1–15. https://doi.org/10.1038/s41598-024-55192-5.

Omer, S. S. & Hassan, N. E. (2024). Application of biodegradable plastic and their environmental impacts: A review, *21*(1), 2139–2148. https://wjarr.com/sites/default/files/WJARR-2024-0155.pdf. https://doi.org/10.30574/WJARR.2024.21.1.0155.

Pannerselvam, A., Kuppusamy, M. S., Shanmugapriyan, J., Kaliappan, V. K. & Sathyamurthy, R. (2022). Experimental investigation of removal of flue gas emissions exhaust from municipal solid waste incinerator using photovoltaic-based electrostatic precipitator. *Environmental Science and Pollution Research, 29*(8), 11209–11218. https://doi.org/10.1007/S11356-021-16378-W/TABLES/5.

Pathak, G., Nichter, M., Hardon, A., Moyer, E., Latkar, A., Simbaya, J., Pakasi, D., Taqueban, E. & Love, J. (2023). Plastic pollution and the open burning of plastic wastes. *Global Environmental Change, 80*, 102648. https://doi.org/10.1016/J.GLOENVCHA.2023.102648.

Payne, J. & Jones, M. D. (2021). The chemical recycling of polyesters for a circular plastics economy: Challenges and emerging opportunities. *ChemSusChem, 14*(19), 4041–4070. https://doi.org/10.1002/CSSC.202100400;REQUESTEDJOURNAL:JOURNAL:1864564X;WGROUP:STRING:PUBLICATION.

Phelan, A. (Anya), Meissner, K., Humphrey, J. & Ross, H. (2022). Plastic pollution and packaging: Corporate commitments and actions from the food and beverage sector. *Journal of Cleaner Production, 331*, 129827. https://doi.org/10.1016/J.JCLEPRO.2021.129827.

Pijnenburg, A. M., Everts, J. W., de Boer, J. & Boon, J. P. (1995). Polybrominated biphenyl and diphenylether flame retardants: analysis, toxicity, and environmental occurrence. *Reviews of Environmental Contamination and Toxicology, 141*, 1–26. https://doi.org/10.1007/978-1-4612-2530-0_1.

Proshad, R., Kormoker, T., Islam, Md. S., Haque, M. A., Rahman, Md. M. & Mithu, Md. M. R. (2017). Toxic effects of plastic on human health and environment : A consequences of health risk assessment in Bangladesh. *International Journal of Health, 6*(1), 1–5. https://doi.org/10.14419/ijh.v6i1.8655.

Qian, K., Wang, L., Teng, J. & Liu, G. (2025). Strategies and technologies for sustainable plastic waste treatment and recycling. *Environmental Functional Materials* [Preprint]. https://doi.org/10.1016/J.EFMAT.2025.01.004.

Qureshi, M. S., Oasmaa, A., Pihkola, H., Deviatkin, I., Tenhunen, A., Mannila, J., Minkkinen, H., Pohjakallio, M. & Laine-Ylijoki, J. (2020). Pyrolysis of plastic waste: Opportunities and challenges. *Journal of Analytical and Applied Pyrolysis, 152*, 104804. https://doi.org/10.1016/J.JAAP.2020.104804.

Rabiu, M. K. & Jaeger-Erben, M. (2024). Reducing single-use plastic in everyday social practices: Insights from a living lab experiment. *Resources, Conservation and Recycling, 200*, 107303. https://doi.org/10.1016/J.RESCONREC.2023.107303.

Rafey, A. & Siddiqui, F. Z. (2023). A review of plastic waste management in India–challenges and opportunities. *International Journal of Environmental Analytical Chemistry, 103*(16), 3971–3987. https://doi.org/10.1080/03067319.2021.1917560;JOURNAL:JOURNAL:GEAC20;REQUESTEDJOURNAL:JOURNAL:GEAC20;PAGE:STRING:ARTICLE/CHAPTER.

Rahman, I. U., Mohammed, H. J. & Bamasag, A. (2025). An exploration of recent waste-to-energy advancements for optimal solid waste management. *Discover Chemical Engineering, 5*(1), 1–26. https://doi.org/10.1007/S43938-025-00079-8.

Rajan, K., Khudsar, F. A. & Kumar, R. (2023). Urbanization and population resources affect microplastic concentration in surface water of the River Ganga. *Journal of Hazardous Materials Advances, 11*, 100342. https://doi.org/10.1016/J.HAZADV.2023.100342.

Rajmohan, K. V. S., Ramya, C., Raja Viswanathan, M. & Varjani, S. (2019). Plastic pollutants: Effective waste management for pollution control and abatement. *Current Opinion in Environmental Science & Health, 12*, 72–84. https://doi.org/10.1016/j.coesh.2019.08.006.

Ramos, A. (2024). Sustainability assessment in waste management: An exploratory study of the social perspective in waste-to-energy cases. *Journal of Cleaner Production, 475*, 143693. https://doi.org/10.1016/J.JCLEPRO.2024.143693.

Reemtsma, T., Weiss, S., Mueller, J., Petrovic, M., González, S., Barcelo, D., Ventura, F. & Knepper, T. P. (2006). Polar pollutants entry into the water cycle by municipal wastewater: A European perspective. *Environmental Science and Technology, 40*(17), 5451–5458. https://doi.org/10.1021/ES060908A/SUPPL_FILE/ES060908ASI20060414_103254.PDF.

Rochman, C. M., Hoh, E., Kurobe, T. & Teh, S. J. (2013). Ingested plastic transfers hazardous chemicals to fish and induces hepatic stress. *Scientific Reports, 3*(1), 3263. https://doi.org/10.1038/srep03263.

Rosenboom, J. G., Langer, R. & Traverso, G. (2022). Bioplastics for a circular economy. *Nature Reviews Materials, 7*(2), 117–137. https://doi.org/10.1038/s41578-021-00407-8.

Rubio, S., Ramos, T. R. P., Leitão, M. M. R. & Barbosa-Povoa, A. P. (2019). Effectiveness of extended producer responsibility policies implementation: The case of Portuguese and Spanish packaging waste systems. *Journal of Cleaner Production, 210*, 217–230. https://doi.org/10.1016/J.JCLEPRO.2018.10.299.

Ryan, P. G., Connell, A. D. & Gardner, B. D. (1988). Plastic ingestion and PCBs in seabirds: Is there a relationship?. *Marine Pollution Bulletin, 19*(4), 174–176. https://doi.org/10.1016/0025-326X(88)90674-1.

Saad, J. M. & Williams, P. T. (2016). Pyrolysis-catalytic-dry reforming of waste plastics and mixed waste plastics for syngas production. *Energy and Fuels, 30*(4), 3198–3204. https://doi.org/10.1021/ACS.ENERGYFUELS.5B02508/ASSET/IMAGES/LARGE/EF-2015-02508J_0005.JPEG.

vom Saal, F. S., Akingbemi, B. T., Belcher, S. M., Birnbaum, L. S., Crain, D. A., Eriksen, M., Farabollini, F., Guillette, L. J., Hauser, R., Heindel, J. J., Ho, S.-M., Hunt, P. A., Iguchi, T., Jobling, S., Kanno, J., Keri, R. A., Knudsen, K. E., Laufer, H., LeBlanc, G. A., Marcus, M., McLachlan, J. A., Myers, J. P., Nadal, A., Newbold, R. R., Olea, N., Prins, G. S., Richter, C. A., Rubin, B. S., Sonnenschein, C., Soto, A. M., Talsness, C. E., Vandenbergh, J. G., Vandenberg, L. N., Walser-Kuntz, D. R., Watson, C. S., Welshons, W. V., Wetherill, Y. & Zoeller, R. T. (2007). Chapel Hill bisphenol A expert panel consensus statement: Integration of mechanisms, effects in animals and potential to impact human health at current levels of exposure. *Reproductive Toxicology, 24*(2), 131–138. https://doi.org/10.1016/j.reprotox.2007.07.005.

Saebea, D., Ruengrit, P., Arpornwichanop, A. & Patcharavorachot, Y. (2020). Gasification of plastic waste for synthesis gas production. *Energy Reports, 6*, 202–207. https://doi.org/10.1016/J.EGYR.2019.08.043.

Saif, S., Razia, E. T., Khan, M., Hatshan, M. R. & Farooq Adil, S. (2025). A comprehensive introduction to solid waste issues. *ACS Symposium Series* [Preprint]. https://doi.org/10.1021/BK-2025-1494.CH001

Samir, A., Ashour, F. H., Hakim, A. A. A. & Bassyouni, M. (2022). Recent advances in biodegradable polymers for sustainable applications. *npj Materials Degradation, 6*(1), 1–28. https://doi.org/10.1038/s41529-022-00277-7.

Santillo, D. & Johnston, P. (2003). Playing with fire: the global threat presented by brominated flame retardants justifies urgent substitution. *Environment International, 29*(6), 725–734. https://doi.org/10.1016/S0160-4120(03)00115-6.

Santos, P., Byrne, A., Karaca, F., Villoria, P., Rio, M. & Pineda-martos, R. (2025). Circular material usage strategies—Principles, circular economy design and management in the built environment. L. Bragança, et al. (ed.) Springer, Cham. https://doi.org/10.1007/978-3-031-73490-8.

Sarkar, D. J., Sarkar, S. Das, V., S. K., Kundu, S. & Das, B. K. (2022). Microplastic pollution in ganga: Present status and future need. *Science and Culture, 88*(November-December). https://doi.org/10.36094/SC.V88.2022.MICROPLASTIC_POLLUTION_IN_GANGA.SARKAR.400.

Sarkingobir, Y., Bello, M. A. & Yabo, H. M. (2021). Harmful effects of plastics on air quality. *Academia Letters* [Preprint]. https://doi.org/10.20935/AL2967.

Seewoo, B. J., Wong, E. V. S., Mulders, Y. R., Goodes, L. M., Eroglu, E., Brunner, M., Gozt, A., Toshniwal, P., Symeonides, C. & Dunlop, S. A. (2024). Impacts associated with the plastic polymers polycarbonate, polystyrene, polyvinyl chloride, and polybutadiene across their life cycle: A review. *Heliyon, 10*(12). https://doi.org/10.1016/J.HELIYON.2024.E32912.

Shahzabeen, A., Ghosh, A., Pandey, B. & Shekhar, S. (2023). Circular economy and sustainable production and consumption, pp 43–65. https://doi.org/10.1007/978-3-031-40304-0_3.

Shen, J., Liang, J., Lin, X., Lin, H., Yu, J. & Wang, S. (2021). The flame-retardant mechanisms and preparation of polymer composites and their potential application in construction engineering. *Polymers, 14*(1), 82. https://doi.org/10.3390/POLYM14010082.

Simantiris, N. (2024). Single-use plastic or paper products? A dilemma that requires societal change. *Cleaner Waste Systems, 7*, 100128. https://doi.org/10.1016/J.CLWAS.2023.100128.

Singh, N. & Walker, T. R. (2024). Plastic recycling: A panacea or environmental pollution problem. *npj Materials Sustainability, 2*(1), 1–7. https://doi.org/10.1038/s44296-024-00024-w.

Singh, P. K., Singh, A., Srivastava, A. K., Chauhan, R., Basniwal, R. K. & Chauhan, A. (2025). Microplastic pollution in the ganga river: A state-of-the-art review of pathways, mechanisms, and mitigation. *Water Supply, 25*(2), 249–267. https://doi.org/10.2166/WS.2025.009.

Sohel Parvez, M., Ullah, H., Faruk, Omar, Simon, Edina, Czédli, Herta, Parvez, M., Juhász-Nagy, P., Faruk, O., Simon, E. & Czédli, H. (2024). Role of microplastics in global warming and climate change: A review. *Water, Air, & Soil Pollution, 235*(3), 1–31. https://doi.org/10.1007/S11270-024-07003-W.

de Souza, F. M. & Gupta, R. K. (2024). Bacteria for bioplastics: progress, applications, and challenges. *ACS Omega, 9*(8), 8666–8686. https://doi.org/10.1021/ACSOMEGA.3C07372/ASSET/IMAGES/LARGE/AO3C07372_0009.JPEG.

Suriapparao, D. V., Kumar, D. A. & Vinu, R. (2022). Microwave co-pyrolysis of PET bottle waste and rice husk: effect of plastic waste loading on product formation. *Sustainable Energy Technologies and Assessments, 49*, 101781. https://doi.org/10.1016/J.SETA.2021.101781.

Tan, W., Cui, D. & Xi, B. (2021). Moving policy and regulation forward for single-use plastic alternatives. *Frontiers of Environmental Science and Engineering, 15*(3), 1–4. https://doi.org/10.1007/S11783-021-1423-5/METRICS.

Taylor, R. L. C. (2019). Bag leakage: The effect of disposable carryout bag regulations on unregulated bags. *Journal of Environmental Economics and Management, 93*, 254–271. https://doi.org/10.1016/J.JEEM.2019.01.001.

Tennakoon, P., Chandika, P., Yi, M. & Jung, W. K. (2023). Marine-derived biopolymers as potential bioplastics, an eco-friendly alternative. *iScience, 26*(4), 106404. https://doi.org/10.1016/J.ISCI.2023.106404.

Tumu, K., Vorst, K. & Curtzwiler, G. (2023). Global plastic waste recycling and extended producer responsibility laws. *Journal of Environmental Management, 348*, 119242. https://doi.org/10.1016/J.JENVMAN.2023.119242.

Vandenberg, L. N., Maffini, M. V., Sonnenschein, C., Rubin, B. S. & Soto, A. M. (2009). Bisphenol-A and the great divide: A review of controversies in the field of endocrine disruption. *Endocrine Reviews, 30*(1), 75–95. https://doi.org/10.1210/er.2008-0021.

Vattanasit, U., Kongpran, J. & Ikeda, A. (2023). Airborne microplastics: A narrative review of potential effects on the human respiratory system. *Science of The Total Environment, 904*, 166745. https://doi.org/10.1016/J.SCITOTENV.2023.166745.

Velis, C. A. & Cook, E. (2021) Mismanagement of plastic waste through open burning with emphasis on the global south: A systematic review of risks to occupational and public health. *Environmental Science and Technology, 55*(11), 7186–7207. https://doi.org/10.1021/ACS.EST.0C08536/ASSET/IMAGES/LARGE/ES0C08536_0002.JPEG.

Verma, R., Vinoda, K. S., Papireddy, M. & Gowda, A. N. S. (2016). Toxic pollutants from plastic waste- A review. *Procedia Environmental Sciences, 35*, 701–708. https://doi.org/10.1016/J.PROENV.2016.07.069.

Villarrubia-Gómez, P., Carney Almroth, B., Eriksen, M., Ryberg, M. & Cornell, S. E. (2024). Plastics pollution exacerbates the impacts of all planetary boundaries. *One Earth, 7*(12), 2119–2138. https://doi.org/10.1016/J.ONEEAR.2024.10.017.

Vollmer, I., Jenks, M. J. F., Roelands, M. C. P., White, R. J., van Harmelen, T., de Wild, P., van der Laan, G. P., Meirer, F., Keurentjes, J. T. F. & Weckhuysen, B. M. (2020). Beyond mechanical recycling: giving new life to plastic waste. *Angewandte Chemie—International Edition, 59*(36), 15402–15423. https://doi.org/10.1002/ANIE.201915651;PAGE:STRING:ARTICLE/CHAPTER.

Vukicevic, A. M., Petrovic, M., Jurisevic, N., Djapan, M., Knezevic, N., Novakovic, A. & Jovanovic, K. (2025). Versatile waste sorting in small batch and flexible manufacturing industries using deep learning techniques. *Scientific Reports, 15*(1), 1–11. https://doi.org/10.1038/s41598-025-87226-x.

Wagner, T. P. (2017). Reducing single-use plastic shopping bags in the USA. *Waste Management, 70*, 3–12. https://doi.org/10.1016/J.WASMAN.2017.09.003.

Wang, Z., Huo, J. & Duan, Y. (2020). The impact of government incentives and penalties on willingness to recycle plastic waste: An evolutionary game theory perspective. *Frontiers of Environmental Science and Engineering, 14*(2), 1–12. https://doi.org/10.1007/S11783-019-1208-2/METRICS.

Wang, Y., Huang, J., Zhu, F. & Zhou, S. (2021). Airborne microplastics: A review on the occurrence, migration and risks to humans. *Bulletin of Environmental Contamination and Toxicology, 107*(4), 657–664. https://doi.org/10.1007/S00128-021-03180-0/TABLES/1.

Waters, Y. L., Wilson, K. A. & Dean, A. J. (2023). Plastic action or distraction? Marine plastic campaigns influence public engagement with climate change in both general and engaged audiences. *Marine Policy, 152*, 105580. https://doi.org/10.1016/J.MARPOL.2023.105580.

Yang, J., Zhao, Y., Li, M., Du, M., Li, X. & Li, Y. (2019). A review of a class of emerging contaminants: The classification, distribution, intensity of consumption, synthesis routes, environmental effects and expectation of pollution abatement to organophosphate flame retardants (OPFRs).

International Journal of Molecular Sciences, 20(12), 2874. https://doi.org/10.3390/IJMS20122874.

Yang, L., Yin, Z., Tian, Y., Liu, Y., Feng, L., Ge, H., Du, Z. & Zhang, L. (2022). A new and systematic review on the efficiency and mechanism of different techniques for OPFRs removal from aqueous environments. *Journal of Hazardous Materials, 431*, 128517. https://doi.org/10.1016/J.JHAZMAT.2022.128517.

Yao, X., Luo, X.S., Fan, J., Zhang, T., Li, H. & Wei, Y. (2022). Ecological and human health risks of atmospheric microplastics (MPs): A review. *Environmental Science: Atmospheres, 2*(5), 921–942. https://doi.org/10.1039/D2EA00041E.

Yin, H., Tang, Z., Meng, T. & Zhang, M. (2020). Concentration profile, spatial distributions and temporal trends of polybrominated diphenyl ethers in sediments across China: Implications for risk assessment. *Ecotoxicology and Environmental Safety, 206*, 111205. https://doi.org/10.1016/J.ECOENV.2020.111205.

Zhang, M., Buekens, A. & Li, X. (2017). Open burning as a source of dioxins. *Critical Reviews in Environmental Science and Technology, 47*(8), 543–620. https://doi.org/10.1080/10643389.2017.1320154.

Zubris, K. A. V. & Richards, B. K. (2005). Synthetic fibers as an indicator of land application of sludge. *Environmental Pollution, 138*(2), 201–211. https://doi.org/10.1016/j.envpol.2005.04.013.

2 Current Solutions for Sustainable Ecosystems

2.1 Introduction

Sustainable ecosystems are those that remain productive and resilient over time, even in the face of external pressures such as pollution, climate change, and human activities (Weiskopf et al., 2020). These ecosystems play a crucial role in conserving biodiversity, maintaining air and water quality, and regulating the global climate. However, with increasing industrialization, exploitation of natural resources, and rapid urban expansion, the sustainability of many ecosystems is under severe threat (Wang & Azam, 2024).

This chapter explores the key environmental challenges that hinder the development and maintenance of sustainable ecosystems, emphasizing the urgent need for effective and innovative solutions.

2.2 Major Challenges to Sustainable Ecosystems

2.2.1 Pollution

Environmental pollution is a pervasive global issue that impacts even the most remote areas—from the poles to the depths of the oceans. Human-driven activities such as industrialization, urbanization, and deforestation have exacerbated pollution levels, deteriorating ecosystems and heightening health risks, particularly in developing countries where 92% of pollution-related deaths occur (Arora et al., 2018).

The economic toll is equally alarming, with annual losses reaching $ 4.6 trillion, equivalent to 6.2% of global GDP (Bhandari, 2019). Air pollution alone has inflated healthcare

N. T. Hatvate et al., *Plastic Waste Management*, Synthesis Lectures on Sustainable Development, https://doi.org/10.1007/978-3-031-96660-6_2

Table 2.1 Types of pollution associated with plastic and their toxic byproducts

Pollution type	Polymer used	Toxic chemicals	Reference
Marine pollution	Polyethylene terephthalate (PET)	Microplastics	(Arora et al., 2018)
Soil/water contamination	Polyvinyl chloride (PVC)	Lead, Cadmium	(Kibria et al., 2023)
E-waste pollution	Polystyrene (PS)	Flame retardants	(Kibria et al., 2023)
Urban pollution	High-density polyethylene (HDPE)	Phthalates	(Kibria et al., 2023)
Industrial pollution	Polypropylene (PP)	Bisphenol A (BPA)	(Kibria et al., 2023)
Air pollution (Incineration)	Mixed plastics	Dioxins, PAHs	(Yue et al., 2023)
Food chain contamination	Nylon	Nanoplastics	(Yue et al., 2023)

costs significantly across both low- and high-income nations. Furthermore, the uncontrolled introduction of over 140,000 synthetic chemicals since 1950, most of which have never undergone toxicity testing, has compounded environmental degradation.

Plastic pollution remains a critical threat. Studies show that 73% of mesopelagic fish in the Northwest Atlantic carry microplastics in their systems, with similar issues affecting over 100 marine species in India, including endangered Olive Ridley Turtles and Gangetic Dolphins (Sivadas et al., 2022). Oil spills, heavy metal contamination, and airborne pollutant deposition have further revealed how far-reaching the impact of pollution truly is. Table 2.1 summarizes the types of pollution associated with plastic and its toxic byproducts.

2.2.2 Climate Change and Global Warming

One of the most visible consequences of environmental degradation is climate change. Since the late nineteenth century, the Earth's surface temperature has risen by approximately 0.74 °C, and projections suggest an increase of 1.4–5.8 °C by the end of the twenty-first century (Bhattacharya, 2019). In 2024, global temperatures had already risen by 1.5 °C compared to the early industrial era average (Ma et al., 2022).

The primary driver is the burning of fossil fuels, coal, oil, and natural gas combined with land-use changes like deforestation and wetland destruction. The effects are wide-ranging: rising sea levels, shrinking Arctic ice, increased frequency of extreme weather events, coral bleaching, and loss of biodiversity. Public health is also impacted, with warmer temperatures altering disease vectors and escalating the risks of illnesses such as dengue and malaria (Weiskopf et al., 2020).

2.2.3 Land Degradation and Agricultural Constraints

Land degradation threatens soil health, food security, and ecosystem stability. Around 25% of the world's land area is degraded, affecting nearly 20% of the global population and resulting in an annual economic loss of approximately $ 230 billion (Gomiero, 2016).

Soil salinization, often caused by poor irrigation practices and low-quality water usage, reduces the fertility of agricultural lands. About 20% of irrigated farmland worldwide suffers from salinity stress, endangering one-third of global food production. Without significant intervention, projections suggest that over 50% of agricultural land could be severely affected by salinization by 2050 (Arora et al., 2018; Kumar et al., 2022).

2.2.4 Habitat and Biodiversity Loss

The loss of biodiversity poses a profound threat to the resilience and functionality of ecosystems. Between 1970 and 2020, global vertebrate population declined by 68%, largely due to human activities such as habitat destruction, pollution, and overexploitation (Weiskopf et al., 2020).

Nearly 420 million hectares of the Earth's forest surface has been modified by human settlement and industrial expansion, putting countless species at risk. Highly biodiverse regions have seen 37,400 species at risk due to land-use changes, with deforestation continuing to erode critical biodiversity hotspots.

2.3 A Complex Issue with Complex Solution

Environmental problems are inherently complex and multi-faceted, arising from a web of interrelated causes that impact ecosystems, human societies, and economies. These challenges are interconnected, often compounding each other's effects, and cannot be addressed in isolation. A holistic and interdisciplinary approach is therefore essential (Kumar et al., 2022).

For example, deforestation reduces carbon sequestration, which accelerates climate change, alters rainfall patterns, and intensifies desertification. Similarly, air pollution contributes to acid rain, degrading soil quality and adversely affecting agricultural productivity (Bhandari, 2019). The United Nations' "Triple Planetary Crisis" framework highlights the critical interconnection between pollution, climate change, and biodiversity loss, emphasizing the need for integrated solutions to safeguard planetary health (Hellweg et al., 2023). Addressing such complex environmental challenges requires coordinated efforts across multiple domains, science, engineering, policymaking, and community action. Strategies such as strengthening waste management systems, enforcing stringent environmental regulations, promoting sustainable agriculture, and transitioning to renewable energy sources have been identified as effective approaches.

The escalating severity of pollution, soil degradation, and biodiversity loss has sparked discussions on optimal restoration strategies. Biological methods, such as microbial treatments, offer efficient and environmentally friendly alternatives to conventional technical solutions. Microorganisms, with their diverse metabolic capabilities, play a critical role in maintaining ecosystem balance by recycling microplastics, breaking down complex polymers, and enhancing soil health (Dixit et al., 2024). Biodegradation, involving the use of microbes to eliminate harmful substances, is widely acknowledged as a cost-effective and sustainable method of pollution reduction.

Traditional siloed approaches have often fallen short, highlighting the need for collaborative frameworks that integrate scientific innovation, policy support, community engagement, and financial incentives. Implementing circular economy principles provides tangible benefits across industries and communities (Dixit et al., 2024; Hossain et al., 2022).

Nature-based solutions (NbS) present another promising approach, leveraging natural processes to tackle environmental challenges such as climate change, flood management, and biodiversity loss (Seddon et al., 2020). By restoring ecosystems, these solutions enhance resilience and offer cost-effective alternatives to traditional infrastructure. Initiatives like coastal mangrove restoration in the Sundarbans of West Bengal, India, exemplify how NbS can simultaneously store carbon and protect against storm surges. Policymakers are increasingly integrating NbS into climate strategies, aligning environmental conservation with economic and social development goals. By implementing NbS, communities can safeguard biodiversity and ecosystem services while strengthening resilience to climate change.

Effective sustainability policies require multi-level governance coordination—from local municipalities to international treaties. Integrated governance systems, such as the European Green Deal, align social, economic, and environmental goals, ensuring a just transition to a carbon–neutral economy (Peng et al., 2024). This initiative encompasses measures for biodiversity preservation, sustainable agriculture, and renewable energy promotion. Policymakers, corporations, and civil society must collaborate to implement frameworks like carbon pricing, green taxes, and sustainable land-use planning. Participatory governance ensures that sustainability policies are practical, inclusive, and capable of addressing local and global environmental challenges (Hellweg et al., 2023).

Artificial Intelligence (AI) and Machine Learning (ML) are revolutionizing natural resource management, offering innovative solutions for environmental conservation. These technologies enable real-time monitoring of water, soil, mineral, biodiversity, and forest resources, ensuring their sustainable use and ethical exploitation. AI models enhance contaminant detection accuracy and optimize remediation strategies, improving the efficiency of environmental management (Peng et al., 2024). Policymakers benefit from AI-driven data analysis, which supports resource availability assessments, identifies critical issues, and aids in decision-making. AI applications extend to climate prediction, pollution control, and disaster response, bolstering resilience and adaptability.

Research across 19 Asian countries from 1990 to 2020 demonstrates that AI-based analysis can help governments forecast carbon emissions based on urbanization, energy consumption, and economic complexity (Peng et al., 2024). The findings underscore AI's potential to advance sustainable industrial practices by highlighting the role of managed resource utilization in reducing CO_2 emissions. AI technologies are also enhancing mineral exploration, water resource management, biodiversity conservation, and energy optimization. For instance, AI-driven multi-indicator analysis has improved the monitoring of plankton communities in marine ecosystems, providing essential insights for sustainable fisheries and marine conservation. In agriculture, AI-enabled precision farming optimizes water usage, minimizes fertilizer application, and monitors soil health, thus boosting food production while minimizing environmental impacts (Kumar et al., 2022).

2.4 Toxic Reduction of Plastic Waste

Plastic waste has become one of the most significant environmental challenges of the twenty-first century. While plastics have revolutionized industries due to their durability, lightweight nature, and cost-effectiveness, their widespread use has led to severe environmental and health concerns (Banerjee & Srivastava, 2012; Kibria et al., 2023). One of the primary issues associated with plastic waste is the presence of toxic substances that persist in the environment, harming ecosystems and human health. Various toxic compounds, including persistent organic pollutants (POPs), heavy metals, and hazardous chemical additives, are released during the production, usage, and disposal of plastics.

2.4.1 Sources and Types of Toxic Substances in Plastic Waste

Plastic waste acts as a reservoir for a broad range of toxic substances that originate from various stages of its lifecycle-from manufacturing and usage to disposal and environmental degradation. The release of these hazardous materials contaminates ecosystems, with wide-ranging impacts on soil, water, air quality, and biodiversity (Sivadas et al., 2022).

2.4.1.1 Heavy Metals

Many plastic products, especially those associated with electronic waste (e-waste), contain heavy metals such as lead, cadmium, chromium, and mercury (Chakraborty et al., 2022). These metals are used to enhance plastic performance, such as stabilizing or adding color. However, when plastics degrade or are improperly disposed of, these toxic metals leach into the environment, contaminating soil and groundwater, and posing serious health hazards, including neurological, renal, and developmental issues (Houessionon et al., 2021).

2.4.1.2 Hazardous Chemical Additives

Several plastic-based products contain chemical additives that enhance their performance but pose environmental and health risks. These include:

- **Phthalates**: Widely used as plasticizers to improve flexibility, especially in PVC products, phthalates are recognized endocrine disruptors that interfere with hormonal systems, potentially causing reproductive and developmental disorders.
- **Bisphenol A (BPA)**: Found in food packaging, water bottles, and consumer goods, BPA is another potent endocrine disruptor linked to cancer, diabetes, and cardiovascular diseases (Kibria et al., 2023).
- **Flame Retardants**: Used in plastics for electronics, textiles, and construction materials, halogenated flame retardants accumulate in living organisms and have been associated with thyroid disruption, immune system suppression, and neurological defects.

2.4.1.3 Microplastics and Nano plastics

The fragmentation of larger plastic debris results in microplastics (<5 mm) and nano plastics (<100 nm), which are now found ubiquitously from the deepest ocean trenches to human bloodstreams. Microplastics not only physically obstruct organisms but also act as vectors, transporting attached pollutants like heavy metals and POPs into biological systems. Their ingestion by marine organisms disrupts physiological functions, and their accumulation in seafood raises concerns about human exposure through the food chain (Kibria et al., 2023; Yue et al., 2023).

2.4.1.4 Persistent Organic Pollutants (POPs)

Plastics absorb and concentrate POPs like PCBs (polychlorinated biphenyls) and DDT from surrounding environments. These highly toxic compounds resist environmental degradation, persist for decades, and bioaccumulate through food chains, leading to severe ecological and human health consequences.

2.4.1.5 Plastic Degradation Products

As plastics degrade, they release harmful compounds such as aldehydes, ketones, and hydrocarbons into the environment. For example, UV-induced photodegradation leads to the release of volatile organic compounds (VOCs), which contribute to ground-level ozone formation, further exacerbating climate change and respiratory illnesses (Barnes et al., 2022).

Table 2.2 summarizes the major components present in plastic waste.

Table 2.2 Major components present in plastic waste

Category	Generation examples
Polyethylene terephthalate (PET)	Water bottle, food jar
Polyvinyl chloride (PVC)	Plumbing pipes, shoe soles
High-density polyethylene (HDPE)	Packaging covers
Polypropylene (PP)	Medicinal waste, chips packet
Polystyrene (PS)	Disposal cups, cutlery, foam
Nylon	Clothing, ropes

2.4.2 Technologies and Methods for Reducing Toxicity in Plastic Waste Management

Advancements in plastic waste management technologies have increasingly focused on reducing toxicity through a range of mechanical, chemical, and biological approaches. Traditional methods often struggle with contamination and inefficiency, but newer innovations seek to eliminate hazardous substances and promote sustainable resource recovery (Hanedan et al.). Effective management now involves a combination of recycling techniques, green material innovation, policy frameworks, and community engagement to minimize the environmental and health risks posed by plastic waste (Banerjee & Srivastava, 2012). Different steps in the recycling of plastics are shown in Fig. 2.1.

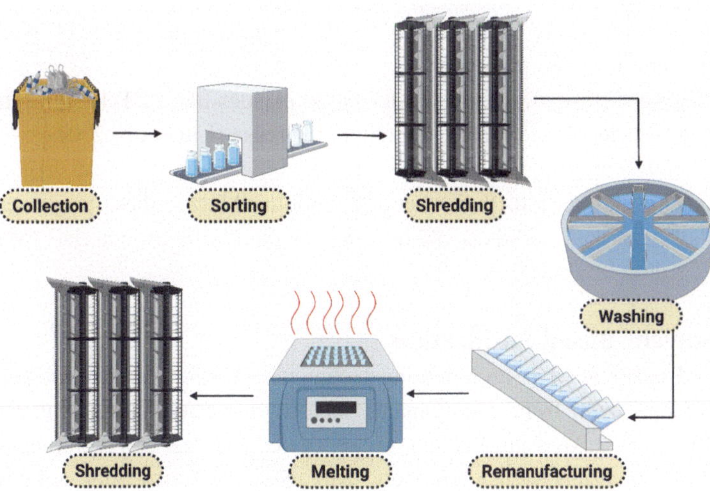

Fig. 2.1 Plastic recycling process (*Created by using Biorender.com*)

2.4.2.1 Mechanical Recycling

Mechanical recycling involves collection, sorting, washing, shredding, and remolding of plastic waste into new products. It is currently one of the most widely adopted recycling methods. However, contamination from toxic additives and degradation of polymer properties during multiple recycling cycles pose significant challenges. Recent advances, such as automated sorting systems equipped with near-infrared (NIR) sensors, robotic separators, and purification technologies, are helping to improve efficiency by accurately identifying plastic types and removing hazardous substances before reprocessing (Kibria et al., 2023; Yue et al., 2023).

Mechanical recycling remains particularly useful for thermoplastics such as PET, HDPE, and polypropylene, but innovations are extending its potential to handle more contaminated or mixed waste streams, thus enhancing the safety and quality of recycled products (Alaghemandi, 2024).

2.4.2.2 Chemical Recycling

Chemical recycling addresses some of the limitations of mechanical recycling by breaking plastics down into their fundamental monomers or basic chemical constituents. This method allows for the production of high-quality recycled materials, even from contaminated or heavily degraded plastic waste. Major chemical recycling technologies include:

- **Pyrolysis**: Involves heating plastic waste in the absence of oxygen to produce synthetic fuels (like pyrolysis oil) and chemical feedstocks. Pyrolysis reduces reliance on fossil fuels and eliminates persistent pollutants (Chang, 2023).
- **Hydrolysis**: Uses water and heat to break down specific types of plastics (e.g., polyesters) into reusable monomers.
- **Glycolysis**: Specifically targets polyester-based plastics like PET, breaking them down using glycols to recover valuable monomers for reuse (Liu et al., 2007).

By decomposing plastics into their chemical building blocks, these processes facilitate the removal of hazardous additives and improve the circularity of materials (Mahajan and Quazi 2017; Kibria et al., 2023).

2.4.2.3 Solvent-Based Purification

Solvent-based purification is an emerging technique wherein selected solvents dissolve plastics to extract them from contaminants such as flame retardants, phthalates, heavy metals, and pigments. After purification, the recovered polymer can be reprecipitated and reprocessed into high-quality new products. This method offers a safer and more precise recycling pathway, especially for complex or multilayer plastic waste streams (Kibria et al., 2023; Yue et al., 2023).

2.4.2.4 Biodegradation and Microbial Treatment

Recent progress in biotechnology has introduced promising methods to tackle plastic waste toxicity. Specific bacterial and fungal strains, such as Ideonella sakaiensis, Pseudomonas, and Aspergillus species, have demonstrated the ability to degrade polymers like PET, polyethylene, and polystyrene into non-toxic byproducts. Engineered microbial enzymes are now being explored to accelerate the breakdown of plastics and reduce the accumulation of microplastics in the environment (Kibria et al., 2023). Biodegradation strategies offer the dual benefit of eliminating plastic pollution while minimizing toxic residuals, presenting an eco-friendly complement to traditional mechanical and chemical recycling approaches (Yue et al., 2023).

2.4.2.5 Case Studies, Global Initiatives and Community-Led Solutions

Several global initiatives illustrate how technological, and policy advancements are reducing toxicity in plastic waste management:

- **European Union**: The EU has implemented stringent bans on single-use plastics and hazardous chemical additives, driving up recycling rates and reducing environmental contamination (Yue et al., 2023)
- **The Netherlands**: Through initiatives promoting circular economy principles, the Netherlands has successfully reduced its reliance on virgin plastics and enhanced reuse and recycling strategies (Banerjee & Srivastava, 2012).
- **Italy**: Plastic valorization projects in Italy focus on transforming waste plastics into high-value materials, encouraging industries to shift toward sustainable production methods (Kibria et al., 2023).
- **Japan**: Japan has pioneered advanced waste-to-energy (WtE) facilities that safely incinerate plastics while capturing toxic emissions through sophisticated filtration systems. Additionally, chemical recycling plants in Japan convert waste plastics into valuable industrial feedstocks, significantly reducing landfilling and mitigating toxicity (Yue et al., 2023; Kibria et al., 2023).

Community-led initiatives in countries like India demonstrate that grassroots action can play a transformative role in plastic waste management. In collaboration with NGOs, waste pickers are engaged in collecting and upcycling plastic waste into construction materials, textiles, and alternative fuels. Such initiatives not only reduce environmental pollution but also create economic opportunities for marginalized communities, highlighting the socio-economic co-benefits of inclusive plastic waste solutions (Banerjee & Srivastava, 2012; Kumar et al., 2022). Figure 2.2 highlights the comprehensive framework for tackling microplastic pollution.

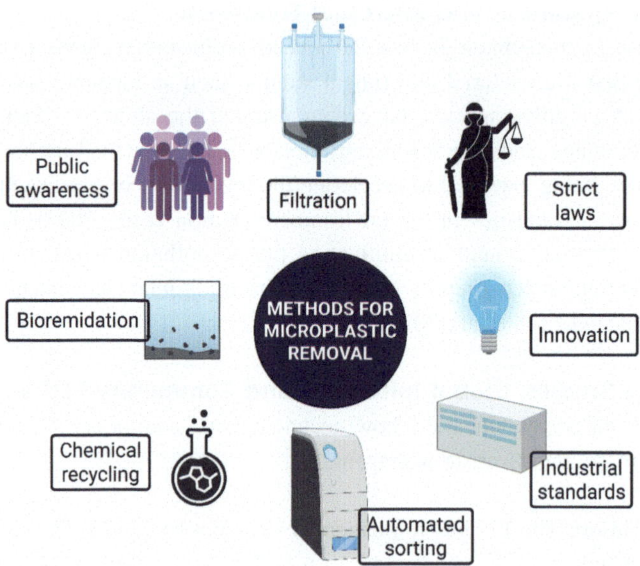

Fig. 2.2 Comprehensive framework for tackling microplastic pollution (*Created by using Biorender.com*)

2.5 Efficiency of Current Plastic Waste Management

2.5.1 Assessment of Existing Plastic Waste Management Systems

Plastic waste management (PWM) systems differ significantly across countries and regions, reflecting disparities in policy enforcement, infrastructure quality, and technological development. In many parts of the world, particularly in developing countries, the lack of a functional waste management infrastructure remains a major hurdle. Reports indicate that only a fraction of plastic waste is properly collected and treated, while the majority ends up in open landfills or is incinerated without adequate emission controls, contributing to severe environmental consequences (Pandey et al., 2023).

In India, for instance, the collection efficiency stands relatively high at approximately 80.28%. However, the actual treatment rate remains low at about 28.4%, primarily due to contamination and insufficient processing facilities (Hossain et al., 2022). The informal sector plays a significant role in waste collection but often operates without safety regulations, exposing workers to serious health risks. Additionally, the lack of investment and proper infrastructure exacerbates inefficiencies in existing waste management systems.

In contrast, developed nations demonstrate relatively better outcomes, driven by stringent environmental regulations and the implementation of structured waste collection and recycling programs. The European Union (EU), for example, has adopted policies promoting Extended Producer Responsibility (EPR) and circular economy principles to boost recycling rates and reduce landfill dependency (Hossain et al., 2022). Nonetheless, even in these regions, challenges persist, particularly with the recycling of complex, multi-layered plastic packaging. Marine and microplastic pollution continue to rise globally, underscoring the urgent need for improved management systems.

2.5.2 Evaluation of Efficiency Metrics and Performance Indicators

Assessing the efficiency of PWM requires the use of key performance indicators (KPIs) such as collection rates, recycling efficiency, landfill diversion rates, and resource recovery levels. One recognized framework is the Environmental Performance Index (EPI), which evaluates countries based on factors such as waste treatment rates, carbon emissions, and pollution control measures (Pandey et al., 2023). A recent study analyzing PWM efficiency in India highlighted that while some regions have adopted advanced waste processing technologies, there is still a heavy reliance on downcycling, where plastics are converted into lower-quality products rather than fully recycled (Hossain et al., 2022).

Standardized data collection methods remain insufficient, complicating efforts to accurately measure progress and formulate effective policies. Furthermore, the presence of non-biodegradable plastics in the waste stream makes recycling more challenging. Research indicates that improving sorting and purification processes can significantly enhance recycling quality and overall waste management efficiency (Hossain et al., 2022). Waste-to-energy (WtE) solutions have been widely debated as a way to boost waste management efficiency. While some experts promote WtE as a viable alternative to landfilling, others raise concerns over the environmental impacts associated with emissions and residue (Kibria et al., 2023).

2.5.3 Challenges and Limitations in Current Practices

Despite notable advancements, several challenges continue to limit the effectiveness of plastic waste management systems. Contamination of waste streams remains a major issue, as it reduces recyclability and increases processing cost. In many countries, the lack of adequate waste segregation infrastructure leads to mixed waste streams that are inefficient and difficult to process. Economic viability is another critical barrier. Virgin plastic production remains cheaper than recycling, primarily due to subsidies favoring fossil fuel industries (Mahajan and Quazi 2017). Policy inadequacies and weak enforcement mechanisms also hamper progress. While initiatives like the EU's EPR regulations show promise, their implementation often varies widely, leading to inconsistent results. The dominance of the informal sector in many developing countries complicates regulatory oversight and endangers workers' health and safety (Hossain et al., 2022).

Moreover, the rapid increase in plastic production continually outpaces recycling and waste management efforts, placing an unsustainable burden on existing systems. Countries such as India, China, and the United States face mounting challenges in managing growing volumes of unmanaged plastic waste due to limited investment in recycling and treatment facilities. Emerging solutions such as biodegradable plastics and chemical recycling offer potential alternatives, but widespread adoption remains limited by high costs, technological hurdles, and infrastructural barriers (Yue et al., 2023).

2.5.4 Future Directions for Improving Plastic Waste Management

Improving the efficiency of plastic waste management systems requires a comprehensive, multi-faceted approach. Firstly, enhancing waste segregation at the source through public education campaigns and policy incentives can significantly improve recycling rates (Kibria et al., 2023). Public participation and community engagement will be vital to the success of source separation initiatives. Secondly, investments in modern recycling technologies, including advanced optical sorting systems, chemical recycling, and the development of biodegradable plastic alternatives, should be prioritized to promote long-term sustainability (Yue et al., 2023).

Policy frameworks should also emphasize the adoption of circular economy models, ensuring continuous reuse and repurposing of plastic materials rather than disposal (Kibria et al., 2023). Encouraging EPR further strengthens this transition by holding manufacturers accountable for the full lifecycle of their products, from design to disposal. In addition, fostering collaborations between governments, private industries, and waste management sectors can lead to better resource mobilization and technological innovation. Public–private partnerships can provide the necessary financial backing for scaling up sustainable waste management solutions. Ultimately, the success of global plastic waste management efforts will depend on the integration of effective regulatory measures, innovative technologies, and widespread public engagement. Transitioning toward a circular economy where plastic waste is minimized, reused, and repurposed efficiently is vital to overcoming the plastic crisis and ensuring a more sustainable future (Hossain et al., 2022).

2.6 Plastic Recycling

Global plastic production has surged dramatically over the past decades. In 2019, annual plastic production reached approximately 460 million metric tons, doubling from 234 million metric tons in 2000. Projections suggest that under a business-as-usual scenario, plastic production could triple to around 1,231 million metric tons by 2060 (Yakubu

Fig. 2.3 7R of recycling process (*Created by using Biorender.com*)

et al., 2024). Despite this rapid increase, global plastic waste management efforts have failed to keep pace. The world is now producing twice as much plastic waste as two decades ago, with the majority of it ending up in landfills, being incinerated, or leaking into the environment. Alarmingly, only about 9% of plastic waste is currently successfully recycled. The 7R recycling processes are illustrated in Fig. 2.3.

It has become increasingly clear that traditional plastic waste management approaches, primarily mechanical recycling and landfilling, are inadequate and unsustainable. Landfilling, the most common disposal method, poses serious environmental hazards (Yue et al., 2023). Plastics, which are highly resistant to degradation, can persist in landfills for centuries, gradually releasing toxic compounds into surrounding soil and groundwater, with detrimental effects on ecosystems and human health. Additionally, the degradation of plastics in landfills contributes to climate change by releasing methane and other greenhouse gases (Alaghemandi, 2024).

Although mechanical recycling is often promoted as a more environmentally friendly alternative, it faces several significant challenges that limit its effectiveness. Contamination, polymer degradation, and limited applicability across different plastic types all reduce the efficiency and quality of recycled materials (Alaghemandi, 2024).

2.6.1 Types and Recyclability of Plastics

Understanding the characteristics and recyclability of different plastics is essential for designing effective waste management strategies.

- Polyethylene Terephthalate (PET) and High-Density Polyethylene (HDPE) are among the most widely recycled plastics, commonly found in beverage bottles and packaging (Muringayil Joseph et al., 2024).
- Low-Density Polyethylene (LDPE), used in plastic bags and films, presents challenges due to contamination and lightweight properties that complicate processing (Jebashalomi, Emmanuel Charles & Rajaram, 2024).
- Polyvinyl Chloride (PVC), found in pipes and vinyl products, has limited recyclability due to the presence of hazardous chemical additives (Lu et al., 2023).
- Polypropylene (PP), used in automotive parts and packaging, and Polystyrene (PS), found in disposable cups and insulation materials, both pose recycling difficulties due to brittleness and low density (Achilias et al., 2007).

Addressing these material-specific challenges is critical to improving recycling rates and ensuring a more sustainable plastics economy.

2.6.2 Environmental Benefits of Plastic Recycling

Recycling plastics offers significant environmental advantages. It can substantially reduce carbon dioxide emissions and energy consumption. By recycling plastics, China's sustainable industrial parks reduce greenhouse gas emissions by 14 million tonnes, which is the same as taking more than 3 million cars off the road (Cudjoe et al., 2021). Additionally, recycling reduces pressure on landfills and helps prevent plastic pollution in oceans and terrestrial ecosystems. It was discovered that the amount of plastic garbage recycled every day worldwide increased from 58.8 thousand tonnes per day in 2012 to 82.0 thousand tonnes per day, a 39.5% increase (Cudjoe et al., 2021).

However, recycling processes are not without environmental drawbacks. Mechanical recycling consumes considerable energy and can release microplastics into the environment during processing stages. These issues underline the need for continued innovation in recycling technologies to mitigate potential negative effects.

2.6.3 Role of Stakeholders in Enhancing Recycling

Improving plastic recycling efficiency requires coordinated participation across multiple levels:

2.6.3.1 Individuals and Households

Proper segregation of waste at the source and cleaning of plastics before disposal significantly enhance the quality and efficiency of recycling. Awareness of recycling codes and material types is vital.

2.6.3.2 Retailers and Businesses

Retailers can promote deposit-refund systems for plastic bottles and prioritize recyclable packaging materials.

2.6.3.3 Industries and Manufacturers

Adopting eco-design principles that prioritize recyclability, implementing closed-loop recycling systems, and incorporating recycled content into new products are essential steps toward sustainability. Chemical recycling technologies are emerging as promising solutions for hard-to-recycle plastics. By breaking down plastics into their original monomers or chemical constituents, chemical recycling enables higher-quality material recovery compared to mechanical methods.

2.6.3.4 Global Best Practices and Corporate Initiatives

Countries excelling in plastic recycling, such as Germany, Sweden, and Japan, provide valuable insights into effective waste management strategies. Germany has established stringent recycling laws and advanced sorting systems, leading to high recycling rates. Sweden has successfully integrated waste-to-energy processes, ensuring minimal landfill usage, while Japan and Thailand enforce strict waste sorting regulations, fostering public participation in recycling programs (Liang et al., 2021).

Governments worldwide are adopting policies such as EPR, which mandates that manufacturers take responsibility for their products' end-of-life disposal (Cao et al., 2016). Plastic taxes, bans on non-recyclable plastics, and subsidies for companies using recycled materials further incentivize sustainable practices. Corporate initiatives, including Coca-Cola's "World Without Waste" program and Unilever's plastic reduction strategies, demonstrate the private sector's role in advancing recycling efforts.

2.6.3.5 Future Trends in Plastic Recycling

The future of plastic recycling lies in technological innovation and stronger policy frameworks.

- AI-driven sorting technologies are revolutionizing recycling plants by improving the precision of material separation and reducing contamination rates (Peng et al., 2024).
- Enzymatic plastic degradation is an emerging biotechnology that enables the breakdown of plastics into reusable base molecules under mild conditions, potentially overcoming the limitations of conventional recycling methods (Alaghemandi, 2024).

- Biodegradable and compostable plastics are gaining attention as alternatives, although challenges remain in ensuring their effective decomposition outside controlled industrial settings.

Transitioning toward a circular economy, where plastics are designed, used, and recycled in closed loops, is critical to mitigating plastic pollution in the long term. Emphasizing sustainable design, material innovation, public education, and robust policy enforcement will be essential to achieving this vision (Hossain et al., 2022). By integrating technological advancement, policy initiatives, and behavioral changes at all levels of society, plastic recycling can significantly reduce environmental harm, conserve natural resources, and support a more sustainable global future. Continued investment in research, international collaboration, and public awareness will be vital to building resilient, efficient, and inclusive plastic waste management systems.

2.7 Conclusion

Addressing environmental sustainability requires a multidimensional strategy integrating waste management, ecosystem restoration, and nature-based solutions. Throughout this chapter, the critical challenges and emerging solutions have been discussed, highlighting the urgent need for systemic change.

Plastic pollution remains one of the most persistent threats to both ecosystems and human health. With only a small fraction of global plastic waste recycled and the majority ending up in landfills or natural environments, traditional disposal methods, landfilling, incineration, and mechanical recycling, have proven inadequate. However, technological advancements such as catalytic recycling, chemical depolymerization, and microbial degradation offer promising pathways for sustainable plastic management.

Alongside waste reduction, ecosystem restoration emerges as a cornerstone for achieving global sustainability goals. Initiatives like reforestation, wetland revival, and regenerative agriculture enhance biodiversity, restore ecological balance, and build resilience against climate change. Nature-based solutions (NbS), leveraging natural processes for environmental recovery, provide cost-effective and adaptive alternatives to conventional infrastructure, although broader implementation faces policy and funding challenges.

Ultimately, achieving environmental sustainability demands a collective effort. innovation, circular economy principles, inclusive policies, and grassroots participation must converge to shift societies toward regenerative development models. By viewing waste as a resource, investing in emerging technologies, and harmonizing economic growth with ecological stewardship, it is possible to restore the balance between humanity and the planet. Through consistent, collaborative action, realizing the vision of ecosystem restoration and sustainable development by 2030 is not only possible-it is imperative.

References

Achilias, D. S., Roupakias, C., Megalokonomos, P., Lappas, A. A. & Antonakou, E. V. (2007). Chemical recycling of plastic wastes made from polyethylene (LDPE and HDPE) and polypropylene (PP). *Journal of Hazardous Materials*, *149*(3), 536–542. https://doi.org/10.1016/j.jhazmat.2007.06.076.

Alaghemandi, M. (2024) Sustainable solutions through innovative plastic waste recycling technologies. *Sustainability*, *16*(23), 10401. https://doi.org/10.3390/su162310401.

Arora, N. K., Fatima, T., Mishra, I., Verma, M., Mishra, J. and Mishra, V. (2018) Environmental sustainability: challenges and viable solutions. *Environmental Sustainability*, *1*(4), 309–340. https://doi.org/10.1007/s42398-018-00038-w.

Banerjee, T. & Srivastava, R. K. (2012) Plastics waste management and resource recovery in India. *International Journal of Environment and Waste Management*, 10(1), 90. https://doi.org/10.1504/IJEWM.2012.048153.

Barnes, P. W., Robson, T. M., Neale, P. J., Williamson, C. E., Zepp, R. G., Madronich, S., Wilson, S. R., Andrady, A. L., Heikkilä, A. M., Bernhard, G. H., Bais, A. F., Neale, R. E., Bornman, J. F., Jansen, M. A. K., Klekociuk, A. R., Martinez-Abaigar, J., Robinson, S. A., Wang, Q.- W., Banaszak, A. T., Häder, D.- P., Hylander, S., Rose, K. C., Wängberg, S.- Å., Foereid, B., Hou, W.- C., Ossola, R., Paul, N. D., Ukpebor, J. E., Andersen, M. P. S., Longstreth, J., Schikowski, T., Solomon, K. R., Sulzberger, B., Bruckman, L. S., Pandey, K. K., White, C. C., Zhu, L., Zhu, M., Aucamp, P. J., Liley, J. B., McKenzie, R. L., Berwick, M., Byrne, S. N., Hollestein, L. M., Lucas, R. M., Olsen, C. M., Rhodes, L. E., Yazar, S. & Young, A. R. (2022) Environmental effects of stratospheric ozone depletion, UV radiation, and interactions with climate change: UNEP Environmental Effects Assessment Panel, Update 2021. *Photochemical & Photobiological Sciences*, *21*(3), 275–301. https://doi.org/10.1007/s43630-022-00176-5.

Bhandari, M. P. (2019). "BashudaivaKutumbakkam"- The entire world is our home and all living beings are our relatives. Why we need to worry about climate change, with reference to pollution problems in the major cities of India, Nepal, Bangladesh and Pakistan. *Advances in Agriculture and Environmental Science: Open Access (AAEOA)*, *2*(1), 8–35. https://doi.org/10.30881/aaeoa.00019.

Bhattacharya, A. (2019). Global climate change and its impact on agriculture. In, *Changing Climate and Resource Use Efficiency in Plants*. Elsevier, pp 1–50. https://doi.org/10.1016/B978-0-12-816209-5.00001-5.

Cao, J., Lu, B., Chen, Y., Zhang, X., Zhai, G., Zhou, G., Jiang, B. & Schnoor, J. L. (2016). Extended producer responsibility system in China improves e-waste recycling: Government policies, enterprise, and public awareness. *Renewable and Sustainable Energy Reviews*, *62*, 882–894. https://doi.org/10.1016/j.rser.2016.04.078.

Chakraborty, S. C., Qamruzzaman, M., Zaman, M. W. U., Alam, M. M., Hossain, M. D., Pramanik, B. K., Nguyen, L. N., Nghiem, L. D., Ahmed, M. F., Zhou, J. L., Mondal, Md. I. H., Hossain, M. A., Johir, M. A. H., Ahmed, M. B., Sithi, J. A., Zargar, M. & Moni, M. A. (2022). Metals in e-waste: Occurrence, fate, impacts and remediation technologies. *Process Safety and Environmental Protection*, *162*, 230–252. https://doi.org/10.1016/j.psep.2022.04.011.

Chang, S. H. (2023). Plastic waste as pyrolysis feedstock for plastic oil production: A review. *Science of The Total Environment*, *877*, 162719. https://doi.org/10.1016/j.scitotenv.2023.162719.

Cudjoe, D., Zhu, B., Nketiah, E., Wang, H., Chen, W. & Qianqian, Y. (2021). The potential energy and environmental benefits of global recyclable resources. *Science of The Total Environment*, *798*, 149258. https://doi.org/10.1016/j.scitotenv.2021.149258.

Dixit, M., Ghoshal, D., Lal Meena, A., Ghasal, P. C., Rai, A. K., Choudhary, J. & Dutta, D. (2024). Changes in soil microbial diversity under present land degradation scenario. *Total Environment Advances*, *10*, 200104. https://doi.org/10.1016/j.teadva.2024.200104.

Gomiero, T. (2016). Soil degradation, land scarcity and food security: Reviewing a complex challenge. *Sustainability*, *8*(3), 281. https://doi.org/10.3390/su8030281.

Hellweg, S., Benetto, E., Huijbregts, M. A. J., Verones, F. & Wood, R. (2023). Life-cycle assessment to guide solutions for the triple planetary crisis. *Nature Reviews Earth & Environment*, *4*(7), 471–486. https://doi.org/10.1038/s43017-023-00449-2.

Hossain, R., Islam, M. T., Shanker, R., Khan, D., Locock, K. E. S., Ghose, A., Schandl, H., Dhodapkar, R. & Sahajwalla, V. (2022). Plastic waste management in india: challenges, opportunities, and roadmap for circular economy. *Sustainability*, *14*(8), 4425. https://doi.org/10.3390/su14084425.

Houessionon, M. G. K., Ouendo, E.-M. D., Bouland, C., Takyi, S. A., Kedote, N. M., Fayomi, B., Fobil, J. N. & Basu, N. (2021). Environmental heavy metal contamination from electronic waste (E-Waste) recycling activities worldwide: A systematic review from 2005 to 2017. *International Journal of Environmental Research and Public Health*, *18*(7), 3517. https://doi.org/10.3390/ijerph18073517.

Jebashalomi, V., Emmanuel Charles, P. & Rajaram, R. (2024). Microbial degradation of low-density polyethylene (LDPE) and polystyrene using Bacillus cereus (OR268710) isolated from plastic-polluted tropical coastal environment. *Science of The Total Environment*, *924*, 171580. https://doi.org/10.1016/j.scitotenv.2024.171580.

Kibria, Md. G., Masuk, N. I., Safayet, R., Nguyen, H. Q. & Mourshed, M. (2023). Plastic waste: challenges and opportunities to mitigate pollution and effective management. *International Journal of Environmental Research*, *17*(1), 20. https://doi.org/10.1007/s41742-023-00507-z.

Kumar, S., Singh, P., Verma, K., Kumar, P. & Yadav, A. (2022). Environmental issues and their possible solutions for sustainable development, India: A review. *Current World Environment*, *17*(3), 531–541. https://doi.org/10.12944/CWE.17.3.3.

Liang, Y., Tan, Q., Song, Q. & Li, J. (2021). An analysis of the plastic waste trade and management in Asia. *Waste Management*, *119*, 242–253. https://doi.org/10.1016/j.wasman.2020.09.049.

Liu, L., Cheng, S. Y., Li, J.B. & Huang, Y. F. (2007). Mitigating environmental pollution and impacts from fossil fuels: the role of alternative fuels. *Energy Sources, Part A: Recovery, Utilization, and Environmental Effects*, *29*(12), 1069–1080. https://doi.org/10.1080/15567030601003627.

Lu, L., Li, W., Cheng, Y. & Liu, M. (2023). Chemical recycling technologies for PVC waste and PVC-containing plastic waste: A review. *Waste Management*, *166*, 245–258. https://doi.org/10.1016/j.wasman.2023.05.012.

Ma, N., Jiang, J. H., Hou, K., Lin, Y., Vu, T., Rosen, P. E., Gu, Y. & Fahy, K. A. (2022) '21st century global and regional surface temperature projections. *Earth and Space Science*, *9*(12). https://doi.org/10.1029/2022EA002662.

Muringayil Joseph, T., Azat, S., Ahmadi, Z., Moini Jazani, O., Esmaeili, A., Kianfar, E., Haponiuk, J. & Thomas, S. (2024). Polyethylene terephthalate (PET) recycling: A review. *Case Studies in Chemical and Environmental Engineering*, *9*, 100673. https://doi.org/10.1016/j.cscee.2024.100673.

Pandey, P., Dhiman, M., Kansal, A. & Subudhi, S. P. (2023). Plastic waste management for sustainable environment: techniques and approaches. *Waste Disposal & Sustainable Energy*, *5*(2), 205–222. https://doi.org/10.1007/s42768-023-00134-6.

Peng, X., Guan, X., Zeng, Y. and Zhang, J. (2024). Artificial intelligence-driven multi-energy optimization: promoting green transition of rural energy planning and sustainable energy economy. *Sustainability*, *16*(10), 4111. https://doi.org/10.3390/su16104111.

Seddon, N., Chausson, A., Berry, P., Girardin, C. A. J., Smith, A., Turner, B. (2020). Understanding the value and limits of nature-based solutions to climate change and other global challenges. *Philosophical Transactions of the Royal Society B Biology Scicence 375*(1794), 20190120. https://doi.org/10.1098/rstb.2019.0120.

Sivadas, S. K., Mishra, P., Kaviarasan, T., Sambandam, M., Dhineka, K., Murthy, M. V. R., Nayak, S., Sivyer, D. and Hoehn, D. (2022). Litter and plastic monitoring in the Indian marine environment: A review of current research, policies, waste management, and a roadmap for multidisciplinary action. *Marine Pollution Bulletin, 176*, 113424. https://doi.org/10.1016/j.marpolbul.2022.113424.

Wang, J. & Azam, W. (2024). Natural resource scarcity, fossil fuel energy consumption, and total greenhouse gas emissions in top emitting countries. *Geoscience Frontiers, 15*(2), 101757. https://doi.org/10.1016/j.gsf.2023.101757.

Weiskopf, S. R., Rubenstein, M. A., Crozier, L. G., Gaichas, S., Griffis, R., Halofsky, J. E., Hyde, K. J. W., Morelli, T. L., Morisette, J. T., Muñoz, R. C., Pershing, A. J., Peterson, D. L., Poudel, R., Staudinger, M. D., Sutton-Grier, A. E., Thompson, L., Vose, J., Weltzin, J. F. & Whyte, K. P. (2020). Climate change effects on biodiversity, ecosystems, ecosystem services, and natural resource management in the United States. *Science of The Total Environment, 733*, 137782. https://doi.org/10.1016/j.scitotenv.2020.137782.

Yakubu, S., Miao, B., Hou, M. & Zhao, Y. (2024). A review of the ecotoxicological status of microplastic pollution in African freshwater systems. *Science of The Total Environment, 946*, 174092. https://doi.org/10.1016/j.scitotenv.2024.174092.

Yue, S., Wang, P., Yu, B., Zhang, T., Zhao, Z., Li, Y. & Zhan, S. (2023). From plastic waste to treasure: selective upcycling through catalytic technologies. *Advanced Energy Materials, 13*(41). https://doi.org/10.1002/aenm.202302008.

Innovative Approaches to Plastic Waste Management

3

3.1 Introduction

Since the mid-twentieth century, global plastic production has surged, driven by its exceptional durability, versatility, and cost-effectiveness. Its lightweight nature, affordability, and adaptability have made plastic an indispensable material across industries such as packaging, healthcare, electronics, construction, and transportation (UNEP, 2024). However, the widespread and uncontrolled use of plastics has led to a severe environmental crisis, with annual production exceeding 400 million tons. The improper disposal of plastic waste (PW) has resulted in persistent pollution, microplastic contamination, soil degradation, marine ecosystem destruction, and significant threats to both biodiversity and human health.

The end-of-life management of plastic waste has become a pressing global challenge, as traditional disposal methods like landfilling and incineration contribute to pollution, greenhouse gas emissions (CO_2, methane, and toxic fumes such as dioxins and furans), and long-term ecological damage (Jambeck et al., 2015). Studies indicate that nearly 11 million metric tons of plastic enter the ocean annually, disrupting marine ecosystems and leading to bioaccumulation of microplastics in seafood consumed by humans (Lebreton et al., 2017). These challenges underscore the urgent need for innovative, sustainable, and circular solutions to mitigate the environmental impact of plastic waste.

To address this escalating crisis, governments, industries, and researchers are actively exploring advanced waste management strategies, including cutting-edge recycling technologies, biodegradable alternatives, biotechnological solutions (microbial and enzymatic plastic degradation), and circular economy frameworks (Ellen Macarthur Foundation; World Economic Forum; McKinsey & Co., 2016). The circular economy approach emphasizes reducing plastic production, improving waste collection, and creating closed-loop recycling systems to minimize environmental footprints (Ellen Macarthur Foundation;

N. T. Hatvate et al., *Plastic Waste Management*, Synthesis Lectures on Sustainable Development, https://doi.org/10.1007/978-3-031-96660-6_3

World Economic Forum; McKinsey & Co., 2016). Innovations such as chemical recycling, pyrolysis, and AI-driven waste sorting are revolutionizing the industry by enhancing material recovery and minimizing downcycling losses (Hopewell et al., 2009).

This chapter delves into these approaches by analyzing successful local experiences, emerging waste treatment technologies, and pollution control mechanisms. Through real-world case studies, technological advancements, policy interventions, and community-driven initiatives, this work explores how various stakeholders contribute to tackling the plastic waste crisis. By integrating scientific innovations, industrial best practices, and government regulations, this chapter provides a holistic perspective on the future of plastic waste management, offering scalable and sustainable solutions to minimize plastic waste, mitigate environmental risks, and support global sustainability goals such as those outlined in the United Nations Sustainable Development Goals (SDGs 12, 13, and 14) (Lee et al., 2016).

3.2 Successful Local Experiences in Plastic Waste Management

3.2.1 Community-Led Plastic Waste Initiatives

One of the most impactful approaches to plastic waste management has been the empowerment of local communities through waste collection, segregation, and recycling programs. Several nations, including India, Indonesia, and Brazil, have implemented community-based waste management models that involve informal waste pickers, local businesses, and NGOs (Wilson et al., 2006).

Case Study-1 (Hidayat, 2025)

Batang Kuis District in Indonesia has implemented a Clean Village Program to address waste management challenges and improve environmental quality. The initiative follows a structured approach that includes waste segregation, community engagement, and recycling programs. Key strategies involve educating residents on proper waste disposal, introducing waste banks for recyclable materials, and promoting composting practices. The local government has played a pivotal role by investing in waste collection infrastructure and integrating community participation in waste reduction efforts. As a result, the program has successfully improved waste management practices, reduced pollution, and fostered sustainable environmental habits among residents. Expanding such community-driven initiatives across Indonesia can serve as a model for effective waste management in other regions facing similar challenges.

Case Study-2 (Anon, n.d.)

In Dakar's Bel Air neighborhood, a micro-factory is addressing plastic pollution by converting waste into furniture and paving materials. Initiated by Plastic Odyssey, this project processes five tonnes of plastic waste from the Mbeubeuss landfill into flakes, which are then transformed into boards and pavers. Utilizing low-tech, locally repairable machinery, the factory ensures sustainability and community involvement. Launched on June 4, Plastic Odyssey plans to establish ten such micro-factories across Senegal within two years, with upcoming installations in Saint-Louis and Kaolack. Each unit, operating on a franchise model with local entrepreneurs or municipalities, has the potential to recycle up to 5,000 tonnes of waste annually and create hundreds of jobs. The finished products, including furniture and pavers, are primarily marketed to administrations and large companies, with contracts already in place for various projects, such as the 2026 Youth Olympic Games in Dakar.

3.2.2 Government-Led Policy Interventions

Many countries have adopted a plastic ban, taxation policies, and comprehensive waste management laws to curb plastic pollution at its source. These measures not only restrict single-use plastics but also incentivize eco-friendly alternatives and encourage producer responsibility through Extended Producer Responsibility (EPR) programs. By imposing levies on plastic production and consumption, governments deter excessive plastic use while generating funds for waste management infrastructure. Additionally, stringent enforcement mechanisms, such as penalties for non-compliance and mandatory recycling targets, ensure that plastic waste is systematically managed and diverted from landfills and oceans. When combined with public awareness campaigns and corporate sustainability initiatives, these policies drive a transformative shift toward a circular economy, reinforcing global efforts to mitigate plastic pollution and achieve sustainable development goals (SDGs).

Case Study-1 (Nduwimana et al., 2026)

Rwanda, a small East African nation, has emerged as a global leader in environmental conservation by implementing one of the world's strictest bans on plastic bags and single-use plastics. In 2008, the Rwandan government banned the importation, production, sale, and use of plastic bags, later expanding the law to include additional plastic products in 2019.

The initiative was driven by concerns over environmental pollution, clogged drainage systems, and livestock deaths caused by plastic ingestion. To enforce the ban, the government introduced heavy fines for individuals and businesses found using plastic bags and promoted eco-friendly alternatives such as paper bags and reusable packaging.

One of the most notable impacts of this policy is the transformation of Rwanda's urban landscape. Kigali, the capital city, has been recognized as one of the cleanest cities in Africa, attracting international praise. The policy has also led to job creation in the production of biodegradable alternatives, benefiting small businesses and entrepreneurs.

Despite its success, the ban has faced challenges, particularly from informal traders who rely on plastic packaging. To address this, the government has initiated public awareness campaigns and financial support for businesses transitioning to sustainable packaging. Rwanda's model has since inspired several African countries, including Kenya, Tanzania, and South Africa, to adopt similar regulations.

Case Study-2 (Aragaw, 2025)

Ethiopia faces a growing plastic waste crisis, exacerbating environmental degradation and public health concerns. The country's rapid urbanization and economic growth have led to increased plastic consumption, with limited infrastructure for waste management. To tackle this issue, the Ethiopian government is prioritizing the enforcement of national plastic waste management strategies. Key measures include banning single-use plastics, adopting circular economy principles, and implementing EPR frameworks. By aligning with international best practices, Ethiopia aims to develop a sustainable waste management system that reduces landfill dependency and promotes plastic recycling industries. Investments in green technologies, such as plastic-to-energy solutions and biodegradable alternatives, are also essential for achieving long-term waste reduction and environmental sustainability.

Case Study-3 (Leal Filho et al., 2025)

Marine plastic pollution is a growing global concern, with millions of tons of plastic waste entering the oceans each year. The lack of robust waste management policies and weak enforcement mechanisms significantly contribute to this crisis. A bibliometric analysis of global marine plastic research highlights key areas for intervention, including waste prevention, policy enforcement, and public awareness6. Case studies from various coastal regions demonstrate that countries with strict plastic waste regulations and comprehensive recycling programs have managed to mitigate marine pollution effectively. Recommendations include promoting eco-friendly alternatives, enforcing regulations more stringent on plastic disposal, and fostering global partnerships to address transboundary marine litter. A shift towards a circular economy, coupled with technological advancements in waste treatment, can significantly reduce the environmental burden of marine plastic waste6.

Case Study-4 (Alamgir et al., 2023)

Mozambique, a low-income coastal nation, struggles with inadequate plastic waste management, leading to severe environmental and public health challenges. A life cycle assessment (LCA) study revealed that over 95% of plastic waste in Mozambique ends up in open dumpsites, with around 60% burned, contributing to significant air pollution and health hazards. The study quantified the environmental impacts of plastic waste mismanagement, highlighting its contribution to terrestrial ecotoxicity and climate change. Recommendations for improvement include enhancing waste collection efficiency, developing structured recycling programs, and investing in waste-to-energy

(WtE) technologies. Strengthening policy frameworks and public–private partnerships is also critical to creating a sustainable plastic waste management system in Mozambique.

Case Study-5 (Makarichi et al., 2023)

Zimbabwe faces significant challenges in plastic waste management, with a low collection rate of just 3.7%. A material flow analysis (MFA) study evaluated the country's plastic waste system and modeled different scenarios to assess the impact of various management strategies. The findings suggest that increasing waste collection rates to 10% would enhance the availability of plastic recyclates by 140%, boosting the country's overall recycling rate to 19.8%. However, this alone is insufficient to curb plastic pollution. A combined approach, integrating waste-to-energy (WtE) technologies with improved recycling efforts, is necessary to achieve substantial waste reduction. Implementing these strategies could reduce plastic waste stocks by 40% while generating significant energy resources. The study underscores the importance of a multi-pronged approach to waste management, where policy intervention, recycling infrastructure, and energy recovery solutions work together to achieve sustainable plastic waste disposal.

Case Study-6 (Anon, 2017)

Plastic pollution and deteriorating road infrastructure are two major issues that often seem unrelated. However, in the Netherlands, the Plastic Road initiative has combined these challenges into a single solution by using recycled plastic waste to construct durable and sustainable roads. The idea was first introduced by KWS, Wavin, and TotalEnergies, who developed prefabricated modular road sections made entirely from recycled plastic. These sections are lighter than traditional asphalt, easier to install, and equipped with built-in drainage systems to prevent flooding. A pilot project in the city of Zwolle saw the construction of a bicycle path using Plastic Road technology. The results were promising; the plastic-based roads proved more resilient to heavy rainfall and required less maintenance than traditional roads. Another key benefit is the reduced carbon footprint, as manufacturing plastic roads emits 60% fewer CO_2 emissions compared to conventional asphalt production.

However, challenges remain, such as scalability and ensuring long-term durability. Despite this, the Dutch government is expanding research into this technology, with plans to incorporate plastic roads in highways and urban infrastructure projects. If successful, this innovation could redefine how plastic waste is managed worldwide.

Case Study-7 (Widiastutie et al., 2025)

India generates over 3.5 million tons of plastic waste annually, much of which ends up in rivers and oceans. To address this issue, the government has mandated the use of shredded plastic waste in road construction, a solution that not only manages plastic pollution but also strengthens road durability. The concept was pioneered by Dr. R. Vasudevan, an Indian scientist who discovered that mixing shredded plastic with bitumen creates a stronger and more flexible road surface. The technique has since been officially adopted by the Indian Ministry of Road Transport and Highways, with over 100,000 km of roads already constructed using plastic waste. This initiative has had multiple benefits:

a. Waste Reduction-Reducing the amount of plastic in landfills and the environment.
b. Enhanced Road Durability-Plastic-mixed roads are more resistant to water damage, potholes, and extreme temperatures.
c. Lower Construction Costs-The use of plastic reduces the need for expensive raw materials, making road construction more economical.

Despite its success, the challenge lies in collecting and processing plastic waste efficiently. Currently, efforts are being made to involve local waste pickers and incentivize businesses to contribute to the supply chain. If successfully scaled, this project could be a game-changer for infrastructure development in other developing nations.

3.2.3 Corporate and Industrial Best Practices

Industries and multinational corporations have become key drivers in plastic waste reduction through robust corporate social responsibility (CSR) initiatives, sustainable packaging innovations, and the adoption of closed-loop recycling systems. Recognizing the urgency of the plastic crisis, several corporations have committed to reducing their plastic footprint by investing in biodegradable materials, reusable packaging models, and circular economy practices. Global brands have introduced packaging redesigns, replacing conventional plastics with compostable, recyclable, or refillable alternatives, significantly lowering plastic waste generation. Additionally, industries are actively supporting waste collection and recycling programs, partnering with governments and non-profit organizations to enhance plastic recovery infrastructure. By integrating EPR frameworks, companies are taking ownership of their post-consumer plastic waste, ensuring materials are reintegrated into the production cycle rather than discarded. These proactive measures demonstrate that industry-led innovation, sustainability commitments, and responsible production strategies are essential in achieving long-term solutions to plastic pollution.

Case Study-1 (Nanda et al., 2025)

A promising approach to managing plastic waste involves its incorporation into the construction industry. Researchers have explored the use of high-density polyethylene (HDPE) and polyethylene terephthalate (PET) plastics as replacements for natural coarse aggregates in concrete. The study tested various replacement levels and found that a combination of 10% HDPE and 5% PET yielded the best results in terms of workability, durability, and mechanical strength. This innovative approach not only diverts plastic waste from landfills but also enhances the sustainability of construction materials. The use of plastic waste in concrete production presents a viable alternative to conventional construction materials, reducing environmental impact while promoting circular economy principles. Further research and industrial adoption of plastic-based aggregates could significantly contribute to plastic waste reduction on a global scale.

Case Study-2 (Leadership, n.d.)
As a leading FMCG (Fast-Moving Consumer Goods) company, Unilever has been at the forefront of sustainable packaging initiatives. The company committed to ensuring that all plastic packaging is 100% recyclable, reusable, or compostable by 2025.

Unilever's strategy centers on material innovation with biodegradable alternatives, closed-loop recycling through waste management partnerships, and consumer engagement via education and incentives for responsible disposal.

One of the most notable innovations includes the Cif Ecorefill concept, where concentrated refills are packaged in minimal plastic, reducing packaging waste by 75%. Additionally, Unilever's partnership with Loop (TerraCycle) enables refillable packaging models, significantly reducing single-use plastic.

The initiative has led to:

a. 15% reduction in virgin plastic usage across Unilever's product lines.
b. Increased consumer participation in recycling programs through returnable packaging schemes.
c. Influence on industry-wide sustainability efforts, pushing competitors to adopt similar eco-friendly practices.

3.2.4 Technological Innovations in Waste Management

Technological advancements have revolutionized plastic waste management, significantly improving efficiency, sustainability, and resource recovery. Countries such as Sweden, Germany, and South Korea have embraced cutting-edge innovations like AI-powered waste sorting, advanced chemical recycling, and plastic upcycling technologies, optimizing the entire waste management cycle. Artificial intelligence (AI) and machine learning algorithms have enhanced automated waste segregation, ensuring higher accuracy in identifying and sorting different plastic polymers, thus maximizing recycling potential. Chemical recycling technologies, including pyrolysis and depolymerization, allow plastics to be broken down into their fundamental monomers, enabling the creation of high-quality recycled materials that can replace virgin plastics. Meanwhile, plastic upcycling techniques transform waste plastics into high-value products, including construction materials, textiles, and fuel alternatives. By integrating these innovative approaches, nations are not only minimizing landfill dependence but also fostering a circular economy that promotes sustainable plastic resource utilization and significantly reduces environmental impact.

Case Study-1 (Asia, 2022)

Japan is known for its advanced waste management policies, and when it comes to plastic, the country has developed one of the most efficient recycling systems in the world. Over 85% of Japan's plastic waste is either recycled or converted into energy, setting a benchmark for other nations.

The key to Japan's success lies in its meticulous waste-sorting system, which requires citizens to separate plastics based on type, quality, and recyclability. Unlike conventional recycling, Japan heavily relies on chemical recycling, where plastic waste is broken down into its fundamental monomers, allowing it to be repurposed into high-quality new plastics. Japan's recycling system operates through public–private partnerships, with corporations like Toyota and Panasonic investing in sustainable packaging and plastic waste reduction programs. This has led to innovations such as biodegradable plastics, advanced sorting machines, and AI-driven waste management solutions. While the system is highly effective, it also faces challenges such as high processing costs and the need for consumer compliance. To address this, Japan has launched nationwide awareness campaigns and is investing in next-generation recycling plants that improve efficiency. The country's model demonstrates that with strict regulations, public cooperation, and technological investment, plastic waste can be effectively managed on a national scale.

Case Study-2 (Chertow et al., 2024)

Sweden has one of the most advanced waste management systems in the world, with less than 1% of household waste ending up in landfills. To further optimize plastic recycling, the country has integrated artificial intelligence (AI) and robotics into waste sorting facilities. One of the most successful implementations is the MAX-AI robotic sorting system, which was deployed in the recycling plants in Stockholm and Gothenburg. This AI-powered system utilizes machine learning and computer vision to distinguish between different plastic types, removing contaminants with 99% accuracy. The AI system benefits waste management in several ways:

a. Higher Recycling Efficiency: Faster and more precise sorting increases the amount of high-quality recycled plastics.
b. Reduced Labor Costs: Automating sorting reduces reliance on manual labor, improving safety and efficiency.
c. Data-Driven Optimization: AI collects real-time data to predict plastic waste trends and optimize waste collection routes. As a result, Sweden has increased its plastic recycling rate from 48 to 62% within five years. The success of AI-driven sorting in Sweden has encouraged adoption in other European and North American waste facilities.

3.3 Sustainable Disposal Solutions for Plastic Waste Management

The disposal of plastic waste has become a critical global challenge due to its escalating environmental, ecological, and health impacts. With global plastic production exceeding 400 million tons annually, traditional waste management methods such as landfilling and incineration have proven inadequate, leading to persistent pollution, microplastic contamination, and greenhouse gas emissions (Geyer et al., 2017). Over 79% of all plastic ever produced remains in landfills or the environment, while incineration releases toxic pollutants, worsening air quality and contributing to climate change (Kibria et al., 2023). To address these challenges, innovative and sustainable solutions have emerged, focusing on advanced technologies that enhance waste processing, accelerate plastic breakdown, and optimize resource recovery. AI-driven waste sorting systems now achieve over 95% accuracy in segregating plastic waste, significantly improving efficiency in reuse and repurposing (Chertow et al., 2024). Enzymatic plastic degradation, utilizing specialized enzymes like PETase and MHETase from Ideonella sakaiensis, can break down plastics within days instead of centuries, offering a promising biological approach to plastic waste reduction (Yang et al., 2016). Microwave-assisted depolymerization provides a method for breaking down plastic at the molecular level using high-energy waves, reducing waste volume and facilitating reuse in manufacturing (Alam & Khan, 2024). Nanotechnology is also being explored, with metal oxide nanoparticles like TiO_2 and ZnO acting as photocatalysts to accelerate plastic degradation under sunlight (Akhter et al., 2023), while graphene-based innovations enhance the transformation of discarded plastics into reusable materials (Zafar & Jacob, 2024).

Biotechnological advancements have introduced bacteria and fungi capable of metabolizing plastics, such as Ideonella sakaiensis for PET degradation and Pestalotiopsis microspora, a fungus that decomposes polyurethane efficiently under both aerobic and anaerobic conditions (Yang et al., 2016). Genetic engineering is further refining these biological processes to enhance plastic decomposition rates and adaptability to different environmental conditions. Additionally, automation in waste processing, including AI-powered robotic sorting systems and IoT-enabled smart bins, has revolutionized waste management by increasing processing speed, reducing contamination, and improving material recovery (Chertow et al., 2024). These advancements align with the principles of a circular economy, where plastic waste is continuously repurposed rather than discarded, ensuring long-term sustainability. Scaling up these solutions requires collaborative efforts between researchers, policymakers, and industries to invest in infrastructure, regulatory frameworks, and technological innovations that support efficient and environmentally responsible waste management. As these advancements continue to evolve, they offer a transformative pathway toward mitigating plastic pollution, reducing its ecological footprint, and fostering a cleaner, more sustainable future.

Mechanical recycling continues to be the cornerstone of global plastic recovery efforts. By mechanically sorting, washing, and reprocessing plastics without altering their chemical composition, this method transforms post-consumer waste into secondary raw materials for new products. It is especially effective for relatively clean and homogenous polymers such as polyethylene terephthalate (PET), high-density polyethylene (HDPE), and polypropylene (PP), which can be converted into containers, fibers, and packaging (Singh et al., 2017). However, its effectiveness is constrained by polymer degradation over multiple cycles, contamination by food residues and mixed materials, and limited compatibility between different plastic types. Recent innovations, such as AI-assisted sorting systems, smart labelling, and compatibilizer additives are helping to address these challenges and improve the quality and volume of recyclates (Ragaert et al., 2017).

Chemical recycling, often referred to as advanced recycling, offers a transformative solution to the limitations of mechanical methods. Through techniques like pyrolysis, gasification, depolymerization, and hydrothermal liquefaction, plastics are broken down into their molecular constituents, monomers or syngas, that can be reprocessed into virgin-quality materials or fuels. This makes it particularly valuable for mixed, multilayered, or contaminated plastic waste that mechanical systems cannot process efficiently. Figure 3.1 indicates the plastic pyrolysis pathway for resource and energy recovery.

Although the technology is still maturing, and energy intensity and economic feasibility remain concerns, chemical recycling holds the promise of true circularity without material downgrading. Ongoing investment and collaboration between petrochemical companies,

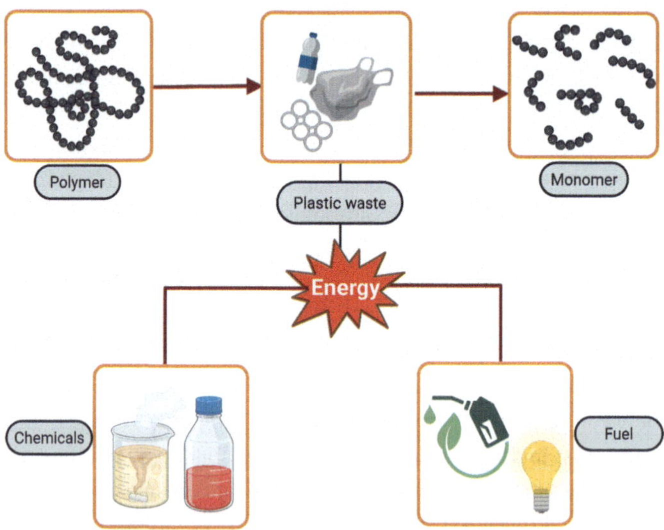

Fig. 3.1 Plastic pyrolysis pathway for resource and energy recovery (*Created by using Biorender.com*)

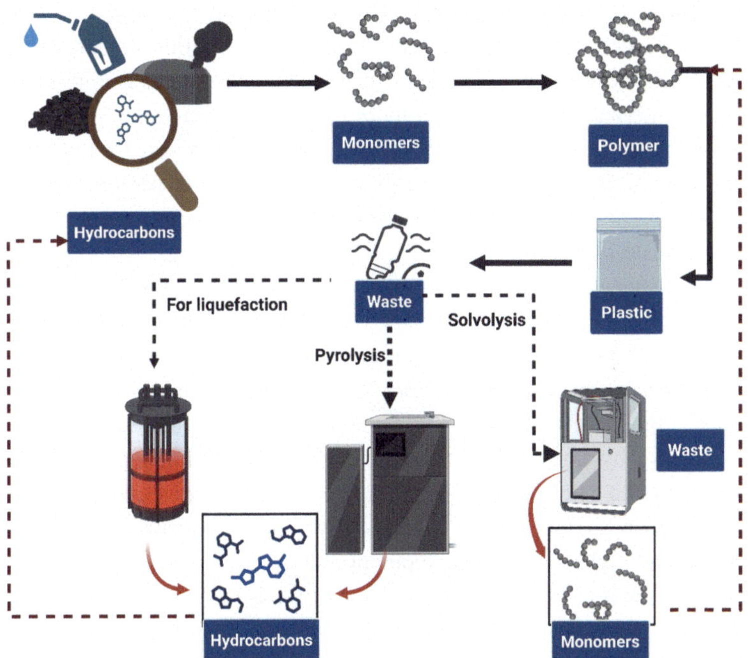

Fig. 3.2 Integrated chemical processes for plastic recycling and upcycling (*Created by using Biorender.com*)

recycling startups, and policymakers are accelerating its commercial viability and regulatory clarity (Kaminsky, 2021; Ragaert et al., 2017; Tewary & Upadhyay 2025). Proposing a 6R framework promoting circular strategies for platform organizations. Digital Policy, Regulation and Governance, 27(2), 145–174.). Figure 3.2 is the integration of chemical processes for plastic recycling and upcycling.

In parallel, biodegradable and compostable plastics are gaining traction as alternatives for specific applications, particularly single-use items like food containers, cutlery, and agricultural films. Derived from renewable sources such as corn starch (PLA) or produced by microbial fermentation (PHA), these polymers are designed to break down into natural components under controlled composting conditions. However, the real-world performance of these materials depends heavily on access to industrial composting infrastructure, which remains unevenly distributed worldwide. Furthermore, consumer confusion and greenwashing pose significant barriers to proper disposal, often leading to contamination of traditional recycling streams. To fully realize their potential, biodegradable plastics must be paired with standardized labeling, targeted education campaigns, and dedicated waste collection systems (Ammendolia & Walker, 2024).

When neither recycling nor composting is viable, waste-to-energy (WTE) systems serve as an important fallback option. Technologies such as incineration with energy recovery, plasma arc gasification, and anaerobic co-digestion convert the calorific value of plastic waste into electricity, heat, or biogas. Modern WTE facilities incorporate advanced emission controls and energy optimization to reduce environmental impact, positioning them as cleaner alternatives to open burning or unmanaged dumping (Arena, 2012). However, WTE should occupy a lower tier in the waste hierarchy, used only when material recovery is impractical. Critics rightly caution that overdependence on incineration may disincentivize reduction and recycling efforts, underscoring the need for clear guidelines and integrated waste planning.

Crucially, the success of all these technological approaches hinges on supportive policy frameworks and infrastructure development. EPR laws, plastic taxes, deposit return schemes, and single-use bans have proven effective in shifting the economic burden of waste management from taxpayers to producers while incentivizing product redesign and sustainable innovation. Nations like Germany, Sweden, and South Korea have set global benchmarks by integrating EPR with well-funded collection, sorting, and recycling systems. In the Global South, community-based initiatives and the inclusion of informal waste workers into formal systems have shown promising results, improving both environmental and socio-economic outcomes.

At the international level, the 2019 amendment to the Basel Convention—which tightened regulations on transboundary movement of plastic waste and ongoing negotiations for a Global Plastics Treaty under the United Nations Environment Assembly (UNEA) represent important steps toward global coordination. These agreements aim to establish common standards for plastic production, disposal, and trade, ensuring transparency, traceability, and environmental justice across borders.

In summary, sustainable plastic disposal requires a multidimensional strategy that harmonizes cutting-edge technologies with sound policy and inclusive governance. Mechanical and chemical recycling, biodegradable materials, waste-to-energy, and regulatory instruments each play a vital role in building a circular and resilient plastic economy. By embracing a holistic, systems-based approach, stakeholders across sectors can turn plastic waste from a global liability into a resource for sustainable development.

3.4 Innovations in Waste Treatment Technologies

In recent decades, plastic has transitioned from a revolutionary material of convenience to one of the most significant environmental challenges facing modern society. Its durability, while once considered an asset, has become its greatest ecological liability (Geyer et al., 2017). Traditional waste disposal methods such as landfilling, open burning, and mechanical recycling are increasingly proving inadequate, both in terms of scale and sustainability. In response, researchers, innovators, and policymakers have turned toward emerging and transformative technologies that aim not merely to manage plastic waste but to reinvent the very paradigm of how plastic is treated at the end of its life cycle.

Explores some of the most promising innovations in waste treatment technologies, focusing on approaches that integrate advanced recycling, nanotechnology, biotechnology, and automation. These technologies not only seek to address the shortcomings of current systems but also aim to create a closed-loop, circular economy where waste becomes a resource and sustainability is embedded in the design and operation of materials.

3.4.1 Advanced Recycling Technologies

At the forefront of innovation is the application of advanced recycling methods that go beyond the mechanical reshaping of plastics. Chemical recycling technologies, such as microwave-assisted depolymerization, gasification enable the breakdown of complex polymer chains into their monomeric forms, which can be repolymerized into virgin-quality plastics (Lopez et al., 2018). This process is particularly effective for mixed or contaminated plastic waste streams that are difficult to manage using conventional techniques. Microwave-assisted depolymerization offers energy efficiency, uniform heat distribution, and faster reaction times, making it a promising alternative to traditional pyrolysis (Kaminsky, 2021).

Artificial intelligence (AI) has also begun to transform sorting processes in material recovery facilities (MRFs). AI-powered systems equipped with high-resolution sensors, machine learning algorithms, and robotic arms can now identify, classify, and sort plastic waste with remarkable precision and speed. These smart-systems reduce contamination, increase material recovery rates, and lower operational costs, paving the way for fully automated recycling plants with minimal human intervention.

Another promising approach lies in enzymatic plastic degradation, which leverages enzymes to selectively hydrolyze synthetic polymers such as polyethylene terephthalate (PET). Engineered enzymes such as PETase, derived from the bacterium Ideonella sakaiensis, have demonstrated the ability to depolymerize PET under mild conditions, offering a low-energy, scalable solution for difficult-to-recycle plastics (Yoshida et al., 2016). This biocatalytic method is being actively refined to increase its reaction rates and substrate range, potentially revolutionizing how polyester-based materials are recycled (Tournier et al., 2020).

3.4.2 Nanotechnology in Plastic Decomposition

Nanotechnology presents transformative opportunities for accelerating plastic degradation through the use of nano-catalysts. These materials, characterized by their high surface area and unique physicochemical properties, can catalyze the breakdown of plastic polymers into benign or recoverable compounds at lower temperatures and shorter reaction times than conventional catalysts (Fotopoulou & Karapanagioti, 2019). Nanoparticles such as titanium dioxide (TiO_2), zinc oxide (ZnO), and iron oxide (Fe_3O_4) have been tested for their photocatalytic activity in degrading polyethylene and polystyrene under UV light, enabling controlled decomposition into carbon dioxide and water (Lee & Li, 2021).

Moreover, nano-structured membranes and filtration systems are being integrated into wastewater treatment plants to capture and break down microplastics before they enter aquatic environments. These nano-enabled interventions offer precision, reusability, and modular scalability, making them ideal for deployment in both urban and decentralized settings (Sun et al., 2019).

3.4.3 Biotechnological Solutions

Among the most promising developments in sustainable plastic waste management are biotechnological approaches that exploit the metabolic capacities of microorganisms. Certain bacteria and fungi have evolved enzymes capable of degrading synthetic polymers, offering an eco-friendly alternative to incineration or landfill disposal.

The bacterium Ideonella sakaiensis, discovered in a Japanese recycling facility in 2016, secretes two enzymes, PETase and MHETase, that can break down PET into its monomers, terephthalic acid and ethylene glycol, which can be reused in plastic production. This discovery has sparked global interest in protein engineering, with researchers working to enhance the catalytic efficiency and substrate specificity of these enzymes (Yoshida et al., 2016).

Similarly, the fungus Pestalotiopsis microspora, isolated from the Amazon rainforest, has been shown to degrade polyurethane, a class of plastic often resistant to traditional recycling, under both aerobic and anaerobic conditions. These microbial processes could be integrated into hybrid treatment systems, combining biological and chemical steps to maximize recovery and minimize environmental impact (Russell et al., 2011).

The real game-changer is the synthetic biology approach, where microbial genomes are edited to enhance their plastic-degrading abilities or to express multi-enzyme complexes capable of digesting various polymers (Danso et al., 2019). Techniques such as CRISPR-Cas9 are being employed to create next-generation microbial consortia that work synergistically to degrade mixed plastic waste (Wei & Zimmermann, 2017).

3.4.4 Automation in Waste Processing

Automation is playing a transformative role in modernizing waste management systems by increasing processing capacity, reducing manual labour, and enhancing safety. Robotics, combined with computer vision and machine learning, is now used extensively in sorting lines to detect and extract different plastic types of based on shape, color, and chemical composition. For example, robotic arms equipped with hyperspectral imaging cameras

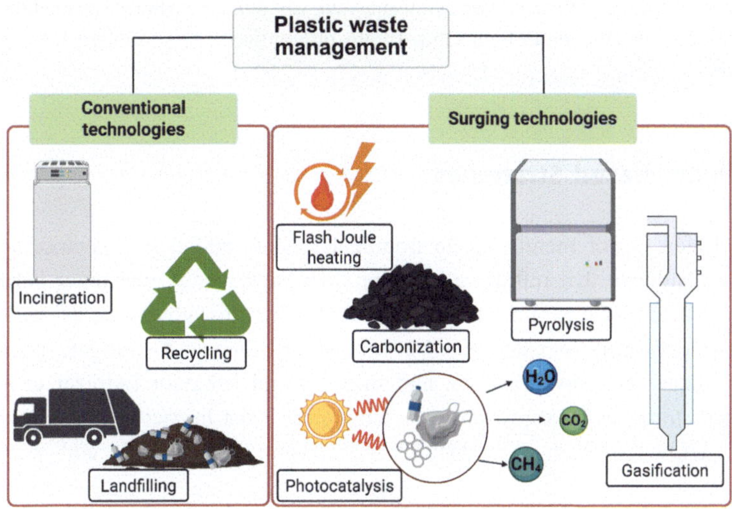

Fig. 3.3 Technological approaches to plastic waste management (*Created by using Biorender.com*)

can distinguish between PET, HDPE, PVC, and other polymers in real-time, ensuring high-purity output streams (Tournier et al., 2020). Fully automated sorting facilities, often termed "Smart MRFs," are capable of operating around the clock with minimal human supervision. These facilities optimize resource recovery and significantly reduce operational inefficiencies. Integration of AI with the Internet of Things (IoT) also allows remote monitoring and adaptive control, further streamlining the recycling workflow (Kaza et al., 2018). Figure 3.3 shows technological approaches to plastic waste management.

3.5 Innovations in Plastic Pollution Control

Plastic pollution has rapidly evolved into one of the most pressing and pervasive environmental challenges of the twenty-first century (Thompson et al., 2009). From urban centres to the most remote corners of the planet, plastic waste is now a ubiquitous contaminant, infiltrating ecosystems, harming wildlife, and posing increasing threats to human health (Term, 2022). Microplastics have been discovered in deep ocean trenches, Arctic ice cores, high-altitude mountain snow, and even within human organs, underscoring the staggering extent of the problem (Li et al., 2020). With over 400 million metric tons of plastic produced annually and an ever-increasing fraction entering our environment, it is clear that conventional clean-up approaches alone are insufficient (Anon, 2022).

Effectively addressing this complex crisis demands a multidimensional strategy—one that integrates bold policy interventions, breakthrough technological innovations, and grassroots community engagement. A transition toward sustainable materials management, circular economic models, and upstream prevention is essential to stem the tide of plastic entering natural systems (Litaudon & Chen, 2023).

3.5.1 Policy-Based Strategies

Plastic pollution is not merely an environmental issue, and it is a socioeconomic and governance challenge that reflects gaps in product stewardship, consumer behavior, and waste infrastructure (Galgani et al., 2019). Recognizing this, governments worldwide are increasingly leveraging legislative and economic instruments to mitigate plastic waste, promote circular economy practices, and shift societal behavior (Alpizar et al., 2020). Policy-based strategies have proven to be among the most impactful levers in this transformation, helping to realign production and consumption systems with sustainability objectives.

This section critically examines leading policy mechanisms, namely bans on single-use plastics, EPR schemes, eco-taxation, and public engagement strategies that are shaping the global trajectory toward a plastic-smart society (Alpizar et al., 2020).

3.5.1.1 Bans on Single-Use Plastics (SUPs)

The global proliferation of single-use plastics (SUPs), such as bags, straws, utensils, and polystyrene containers has made them a primary target of environmental regulation (Kumar et al., 2021). Designed for convenience and immediate disposal, these items are among the most commonly found debris in marine and terrestrial environments. Policy bans on SUPs aim to reduce the volume of non-recyclable waste entering ecosystems and to incentivize the development and adoption of reusable or biodegradable alternatives (Miksch et al., 2022).

European Union (EU)

The EU's Single-Use Plastics Directive, which came into effect on July 3, 2021, represents one of the most ambitious regional frameworks to address SUPs. The directive bans a wide range of plastic products including cutlery, straws, plates, and polystyrene food containers. Furthermore, it mandates that all plastic beverage bottles must contain at least 25% recycled content by 2025 and 30% by 2030. The legislation also includes extended labeling requirements and producer obligations to facilitate proper waste handling (Kasznik & Łapniewska, 2023).

India

India's response to SUPs has been phased and multi-tiered, aligning with its broader waste management and sustainability goals. As of July 1, 2022, the country implemented a nationwide ban on 19 SUP items, such as plastic cutlery, plates, earbuds, and packaging films. This ban is supported by parallel efforts to promote indigenous innovation in sustainable alternatives, including starch-based and compostable bioplastics. Regulatory enforcement is bolstered by state-level implementation strategies and partnerships with industry stakeholders (Chauhan & Sevda, 2023).

Kenya

Kenya made international headlines in 2017 when it enacted one of the world's strictest bans on plastic bags, outlawing their manufacture, sale, and use. While initial compliance challenges were noted, the policy has resulted in dramatic reductions in plastic litter across urban centers, rivers, and wildlife habitats. The success of Kenya's initiative has served as a blueprint for similar legislative efforts across the African continent (Paul & Mironga, 2020).

3.5.1.2 Extended Producer Responsibility (EPR) and Eco-taxation

Beyond restrictions on plastic products, systemic change requires restructuring the incentives that govern how materials are produced, consumed, and discarded. EPR and eco-taxation are two pivotal strategies that internalize the environmental costs of plastic use and create economic signals for sustainable innovation (Flygansvær & Dahlstrom, 2024).

EPR shifts the responsibility for managing post-consumer waste from municipalities to producers, incentivizing companies to design recyclable, resource-efficient products. Germany's Packaging Act exemplifies effective EPR, requiring producers to register, pay material-based fees, and fund national recycling systems, resulting in recovery rates above 60%. France has broadened EPR to cover textiles, electronics, and construction materials through eco-organizations. In Asia, countries like South Korea and Japan implement hybrid EPR systems, combining strict oversight, public–private collaboration, and robust data reporting (Lee & Na, 2010).

Taxation frameworks are driving market shifts toward sustainability. The UK's Plastic Packaging Tax, introduced in 2022, charges £200/ton on packaging with less than 30% recycled content, spurring investment in recycling and alternative materials (Government of the UK, 2023).

3.5.2 Ocean Clean-Up Technologies

The mounting crisis of marine plastic pollution has spurred a wave of technological innovation aimed at both intercepting macroplastic waste and addressing more elusive forms of microplastic contamination. Among the most notable large-scale initiatives is The Ocean Clean-up, a pioneering non-profit organization headquartered in the Netherlands. Leveraging cutting-edge engineering, the project has developed passive, autonomous floating barriers designed to concentrate and extract plastic debris from major oceanic gyres, most prominently the Great Pacific Garbage Patch. These U-shaped systems rely on natural ocean currents to funnel plastics toward a central collection point, where they are retained and periodically retrieved for recycling. Recognizing the importance of upstream intervention, The Ocean Clean-up has also launched solar-powered "Interceptor" vessels in highly polluted rivers, targeting key plastic leakage points in Southeast Asia, Africa, and Latin America. By capturing waste before it reaches marine ecosystems, these scalable, low-energy technologies offer a practical solution to reducing the volume of plastics entering the ocean (Slat, 2023).

In parallel with mechanical recovery systems, biotechnological approaches are gaining traction as complementary solutions for plastic clean-up, particularly for microplastics that are too small or dispersed for physical extraction. Bioremediation, which employs microorganisms capable of breaking down synthetic polymers, has shown considerable promise in laboratory and pilot studies. Microbes such as Ideonella sakaiensis and certain Pseudomonas and Bacillus strains possess enzymatic pathways that can degrade common plastics like polyethylene terephthalate (PET) and polyurethane into environmentally benign byproducts. Fungi like Pestalotiopsis microspora have also demonstrated the ability to metabolize plastic substrates under low-oxygen conditions. Although still in the research phase, these microbial strategies could one day be deployed in marine environments, wastewater treatment facilities, or bioreactors to target persistent microplastic pollutants (Zhou et al., 2022).

Collectively, these ocean clean-up technologies represent a paradigm shift from reactive clean-up efforts to proactive, systems-based interventions. By integrating mechanical engineering with microbial biotechnology, the goal is not only to remove existing plastic debris from aquatic ecosystems but also to create long-term, sustainable solutions that prevent further accumulation. As technological capacity advances and global awareness increases, the fusion of innovation, policy, and environmental stewardship offers a viable pathway to restoring the health of the world's oceans (Matavos-Aramyan, 2024).

3.5.3 Microplastic Filtration Solutions

Microplastics is a plastic particles smaller than 5 mm have become one of the most insidious forms of pollution, infiltrating aquatic ecosystems, terrestrial environments, and even the human food chain. These microscopic fragments originate from diverse sources, including the breakdown of larger plastic debris, synthetic textiles, tire abrasion, and microbeads found in personal care products. Their small size allows them to bypass conventional waste management systems, enabling widespread environmental dispersion and making removal particularly challenging. As research continues to uncover the ecological and health risks posed by microplastics, targeted filtration technologies have emerged as essential tools for mitigating their impact (Coyle et al., 2020).

A critical intervention point is municipal wastewater treatment, where microplastics from households and industry converge. Traditional secondary treatment systems—designed primarily for organic and nutrient removal are not equipped to fully capture these small and buoyant particles. In response, advanced tertiary treatment technologies are being deployed to enhance microplastic retention. Membrane bioreactors (MBRs), ultrafiltration systems, rapid sand filters, and dissolved air flotation (DAF) units are among the most effective methods currently in use, capable of removing up to 99% of microplastic particles. These upgrades not only improve effluent quality but also reduce the potential for microplastic release into natural water bodies, thus providing a vital environmental safeguard (Iyare et al., 2020).

At the household level, growing awareness of synthetic microfiber pollution, primarily released during the washing of synthetic fabrics, has prompted innovation in domestic filtration solutions. Each laundry cycle can release hundreds of thousands of microfibers into wastewater, a substantial contributor to microplastic pollution. In response, appliance manufacturers and environmental policymakers have introduced integrated and aftermarket filters to trap these fibers at the source. France has taken a pioneering step by mandating that all new washing machines sold after 2025 must be equipped with microfiber filters. Meanwhile, external devices such as filter cartridges, laundry bags, and washing machine attachments are gaining popularity among environmentally conscious consumers and are proving effective in significantly reducing fiber discharge into wastewater streams (Erdle et al., 2021).

These filtration solutions, both centralized and decentralized, form a complementary defense against the pervasive spread of microplastics. Their successful deployment hinges not only on technological innovation but also on supportive policy frameworks, industry participation, and consumer education. When implemented at scale, such measures can substantially curb microplastic pollution, preserving water quality, protecting biodiversity, and safeguarding human health. As filtration technologies continue to evolve and regulatory standards become more stringent, microplastic capture will play a pivotal role in building resilient and future-ready waste and water management systems (SAPEA, 2019).

3.6 Plastic in Aquatic Environments

Plastic pollution in aquatic environments has escalated into a global environmental crisis, threatening the health and stability of marine ecosystems, coastal communities, global biodiversity, and public health (Thompson et al., 2009). Its pervasive nature spanning from heavily trafficked coastlines and estuaries to the most remote regions of the world's oceans, underscores the magnitude of the problem. Plastics, owing to their durability, hydrophobicity, and lightweight structure, persist for decades to centuries in marine environments, undergoing fragmentation into microplastics (<5 mm) and nanoplastics

(<100 nm), which further complicate detection, removal, and ecological risk assessment (Li et al., 2020).

According to the United Nations Environment Programme (UNEP), more than 11 million metric tons of plastic waste enter the oceans annually (Marshall, 2018). Without urgent and transformative intervention, this number is projected to nearly triple by 2040, potentially adding over 600 million tons of plastic into marine systems over the next two decades (García-Marín & Rentería, 2024). The primary sources include mismanaged municipal waste, industrial effluents, riverine inputs, abandoned fishing gear, and maritime activities. Once in the ocean, plastics undergo physical, chemical, and biological degradation processes, transforming into persistent microparticles that infiltrate aquatic food chains, disrupt habitat integrity, and absorb toxic pollutants such as persistent organic pollutants (POPs), heavy metals, and endocrine-disrupting compounds (Martín et al., 2022).

The consequences of this pollution are profound. Marine organisms, including plankton, fish, seabirds, turtles, and marine mammals, ingest or become entangled in plastic debris, leading to injury, malnutrition, reduced reproductive success, and mortality (van Emmerik et al., 2022). Furthermore, the ingestion of microplastics by lower trophic species has raised concerns about bioaccumulation and trophic transfer, potentially impacting human health through seafood consumption. Beyond ecological and biological damage, the economic repercussions are equally alarming, with substantial costs attributed to fisheries, aquaculture, tourism, and coastal clean-up efforts (Martín et al., 2022).

This section delves into the complex dynamics of plastic pollution in aquatic environments through a comprehensive and multidisciplinary lens. It outlines the key sources and pathways through which plastic enters marine systems, evaluates its cascading effects on marine life and ecological processes, and reviews global strategies for mitigation, remediation, and prevention. Special emphasis is placed on the roles of international policy frameworks, scientific research, and innovative technologies, including satellite monitoring, filtration systems, and autonomous ocean-cleanup devices, as vital tools in the global response to aquatic plastic pollution. Ultimately, this discussion aims to illuminate the scope of the crisis while highlighting the multifaceted and collaborative efforts required to address it effectively (Rangel-Buitrago et al., 2023).

3.6.1 Sources of Marine Pollution

Marine plastic pollution originates from a wide array of sources, both on land and at sea. These sources vary in scale and nature, from micro-level pollutants like synthetic fibers to large discarded fishing gear, but together, they contribute to a pervasive crisis affecting every corner of the world's aquatic ecosystems. According to UNEP, an estimated 80% of marine plastic debris comes from land-based sources, while the remaining 20% is attributed to ocean-based activities. Understanding the diversity and complexity of these sources is essential for designing effective, targeted interventions (Forrest et al., 2019).

3.6.1.1 Land-Based and Ocean-Based Inputs

Land-based inputs remain the most dominant contributors to plastic pollution in marine environments (Thompson et al., 2009). These include improperly managed municipal solid waste, illegal dumping, stormwater runoff, industrial effluents, and untreated or partially treated sewage. Single-use plastic bags, food wrappers, straws, beverage containers, and takeaway packaging are particularly prevalent, often becoming litter on city streets before making their way into rivers, estuaries, and, ultimately, the ocean (Ali et al., 2022).

Urbanization and inadequate waste infrastructure in rapidly developing regions exacerbate this problem. In many coastal cities, particularly in the Global South, plastic waste is commonly discarded in open dumps or burned in informal settlements, with much of it eventually washed into the sea by wind, rain, or tides (Banu, 2020).

While smaller in volume, ocean-based plastic sources are significantly harmful due to their immediate impact on marine ecosystems (Vergara et al., 2025). These include waste from maritime transport, offshore oil and gas operations, aquaculture, and fishing vessels. Items such as food packaging, ropes, bait containers, and abandoned or lost gear from ships contribute to the growing volume of plastics in marine environments. Coastal tourism further adds to the problem, with beaches and shorelines often littered with bottles, straws, flip-flops, and other disposable items left behind by visitors or swept into the sea (Sinha et al., 2024).

3.6.1.2 Abandoned and Discarded Fishing Gear (Ghost Gear)

Among the most ecologically destructive forms of marine plastic pollution is ghost gear, fishing nets, lines, traps, and other equipment that are lost, discarded, or deliberately abandoned at sea. Constructed from highly durable synthetic polymers like nylon, polypropylene, and polyethylene, ghost gear can drift for decades in the ocean, continuing to entangle marine life in what has been termed "ghost fishing" (Do & Armstrong, 2023).

These nets and lines indiscriminately trap fish, sea turtles, marine mammals, and seabirds, causing injury, starvation, or death. Moreover, ghost gear exacerbates overfishing by depleting commercial fish stocks and damaging coral reefs and seafloor habitats. Economically, it poses a direct threat to coastal livelihoods and small-scale fisheries, especially in low-income regions where communities depend on healthy marine ecosystems (Boussellaa et al., 2024).

Research suggests that ghost gear may comprise as much as 70% of all floating macroplastics in certain regions, particularly near high-intensity fishing zones. The proliferation of ghost gear is fueled by a combination of factors: illegal, unreported, and unregulated (IUU) fishing practices; high costs of retrieval; limited enforcement of marine waste laws; and a lack of economic incentives or return schemes for used equipment (Do & Armstrong, 2023). Programs like the Global Ghost Gear Initiative (GGGI) are now working with governments and industries to retrieve lost gear and develop biodegradable alternatives, but large-scale implementation remains a work in progress (Boussellaa et al., 2024).

3.6.2 Impact on Marine Life and Ecosystems

Plastic pollution exerts multifaceted and often devastating impacts on aquatic life and marine ecosystems. These effects span physical, chemical, biological, and ecological dimensions, compromising organismal health, altering community structures, and weakening ecosystem resilience. From ingestion and entanglement to bioaccumulation and habitat disruption, plastic debris is reshaping the marine environment in profound and sometimes irreversible ways. This section explores the major consequences of plastic waste exposure in marine and coastal systems, drawing on the latest scientific findings to illustrate the magnitude of the threat (Luo et al., 2021).

3.6.2.1 Ingestion and Entanglement

One of the most visible and immediate threats posed by plastic pollution is the ingestion of plastic debris by marine organisms. Countless marine species, including zooplankton, crustaceans, bivalves, fish, sea turtles, seabirds, and cetaceans, mistake plastic fragments for food due to their size, color, shape, or odor. Turtles frequently confuse floating plastic bags for jellyfish, seabirds feed colorful bottle caps to their chicks, and fish ingest microplastics suspended in the water column (Omeyer et al., 2023).

Once ingested, plastics can cause a range of internal issues: mechanical injuries to the gastrointestinal tract, digestive blockages, false satiation (leading to malnutrition), impaired nutrient absorption, and ultimately death. A comprehensive study published in Marine Pollution Bulletin documented over 700 marine species affected by plastic pollution, with ingestion reported in more than 60% of seabird species and every known species of sea turtle (Omeyer et al., 2023).

Equally alarming is the prevalence of entanglement, particularly in ghost gear—abandoned fishing nets, lines, and traps which continues to "ghost fish" long after being discarded. Entangled animals may suffer lacerations, amputations, reduced mobility, and a prolonged death from drowning or starvation. Marine mammals such as seals and dolphins, as well as seabirds and large fish, are especially vulnerable. Plastic rings, straps, and packaging bands can constrict growing organisms, leading to deformities or mortality (Wilcox et al., 2015).

These encounters are not only tragic but are also ecologically and economically significant. Declining populations of affected species disrupt food webs and reduce biodiversity, while commercial fisheries suffer losses from gear entanglement and reduced fish stocks.

3.6.2.2 Toxicological Threats, Bioaccumulation, and Trophic Transfer

Beyond physical harm, plastic pollution presents insidious chemical threats. Plastics are composed of synthetic polymers that often contain additives such as plasticizers (e.g., phthalates), flame retardants, stabilizers, bisphenol A (BPA), and pigments, all of which can leach into surrounding environments (Hermabessiere et al., 2017; Lithner et al., 2011). Moreover, plastic surfaces act as hydrophobic sponges, adsorbing environmental toxins such as polychlorinated biphenyls (PCBs), DDT, and heavy metals from seawater at concentrations far exceeding background levels (Setälä et al., 2014).

When marine organisms ingest plastics, they are exposed not only to the plastic matrix but also to these concentrated toxins (Bakir et al., 2014). Filter feeders like mussels, barnacles, and copepods ingest microplastics from sediments and the water column, which are then transferred up the food chain as predators consume contaminated prey. This process, known as trophic transfer, results in bioaccumulation and biomagnification of harmful chemicals at each successive trophic level (Chae & An, 2017).

The toxicological effects of such exposure can include hormonal disruption, reproductive impairments, behavioral changes, immunotoxicity, and developmental abnormalities94. Notably, these impacts are not confined to aquatic species alone. Human populations that rely heavily on seafood for protein and nutrition may also be at risk, as contaminants enter the human food chain through fish and shellfish consumption (Wright & Kelly, 2017).

Emerging research suggests that chronic exposure to microplastics and associated toxins may compromise human endocrine function, gut health, and immune response, though further studies are needed to quantify long-term health implications (Prata et al., 2020).

3.6.2.3 Coral Reef Degradation and Oxygen Depletion

Plastic pollution is also contributing to the decline of one of the ocean's most critical ecosystems: coral reefs. Coral reef structures built over thousands of years by tiny marine animals are biodiversity hotspots and provide essential services, including shoreline protection, fisheries support, and carbon cycling for an estimated 25% of all marine species despite covering less than 1% of the ocean floor. However, plastic debris poses multiple threats to reef health (Akhtar et al., 2022). Impact of plastic waste on the coral reefs: an overview. *Impact of Plastic Waste on the Marine Biota*, 239–256.).

When plastic settles on coral colonies, it can abrade delicate tissues, introduce pathogens, and block light essential for photosynthesis in the symbiotic algae (zooxanthellae) that corals rely on. According to a study published in Science (2018), the likelihood of coral disease increases from 4 to 89% when plastics are present, primarily due to increased microbial colonization and immune suppression (Lamb et al., 2018).

In addition, plastics disrupt coral reproduction by interfering with larval settlement and growth. Floating debris can also physically smother reef structures, reduce gas exchange, and alter nutrient flows, further weakening reef resilience, especially in the face of climate stressors such as warming and acidification (Lamb et al., 2018).

Similar mechanisms are at play in eutrophic coastal zones, where the accumulation of organic waste and plastics exacerbates oxygen depletion or hypoxia. Plastics trap organic material, which decomposes and consumes dissolved oxygen, creating "dead zones" where most aquatic life cannot survive. These hypoxic conditions result in the loss of benthic fauna, reduced biodiversity, and the collapse of essential ecological functions such as nutrient cycling and sediment stabilization (Vaquer-Sunyer & Duarte, 2008).

3.6.2.4 Ecosystem-Wide Disruption and Habitat Alteration

The impacts of plastic pollution extend beyond individual organisms to entire ecosystems. In benthic environments, plastic fragments alter sediment composition, affecting the diversity and behavior of bottom-dwelling species. Microplastics that settle in sediments can reduce oxygen availability, inhibit microbial processes, and affect bioturbation—the natural mixing of sediment layers by organisms like worms and crabs (Danielli & Sousa, 2025).

Floating plastic debris, including packaging materials and containers, also serves as rafts for invasive species. These "plastic islands" allow non-native organisms to travel long distances across oceans, colonizing new areas and outcompeting native species. Such biological invasions disrupt local food webs and threaten endemic biodiversity, particularly on remote islands and in sensitive marine habitats (Barnes, 2002).

Mangroves and seagrass beds are similarly affected. Plastics accumulate around mangrove roots, restricting gas exchange and water flow. In seagrass meadows, plastic debris can smother plants, reduce light penetration, and hinder seedling establishment—thereby undermining one of the ocean's key carbon-sequestering systems (Bhuyan et al., 2024; Martin et al., 2019).

The cumulative effects of plastic pollution weaken ecosystem resilience, decrease productivity, and reduce the ability of marine environments to provide essential ecosystem services. These disruptions can lead to altered community structures, the collapse of trophic dynamics, and long-term ecological instability (Wali, 2021).

3.6.3 Global Cleanup and Prevention Efforts

As the scale and complexity of plastic pollution in aquatic environments become increasingly apparent, it is equally clear that no single solution can resolve the crisis. Tackling the growing burden of marine plastic requires a multi-pronged, globally coordinated response—one that combines robust international treaties, cutting-edge technological innovations, and local community action. Efforts must address both legacy plastic pollution that has already entered marine systems and the continuous inflow of new plastic waste. This section outlines the evolving global response across three strategic pillars: international agreements and legal frameworks, technological innovations for clean-up, and emerging microplastic filtration solutions (Nikiema & Asiedu, 2022).

3.6.3.1 Technological Innovations in Ocean Clean-Up

While policy plays a foundational role in plastic prevention, technological innovation is vital for addressing the vast quantities of plastic already polluting our oceans. In recent years, scientists, engineers, and entrepreneurs have developed innovative tools and systems to collect, monitor, and remove plastic debris from aquatic environments. These technologies vary in scope and application, from passive systems targeting macroplastics to satellite imaging for pollution mapping (Gkanasos et al., 2021).

Floating Barriers-The Ocean Clean-up: One of the most high-profile initiatives in this space is The Ocean Clean-up, a Dutch non-profit organization that has designed autonomous floating barriers to capture plastic debris from major ocean gyres (Frantzi et al., 2021). Using the natural movement of currents, these U-shaped systems passively corral plastic waste, which is later collected and shipped for recycling. The Ocean Clean-up has successfully deployed these systems in the Great Pacific Garbage Patch and is now scaling operations to address similar hotspots worldwide (Bellou et al., 2021).

River Interceptors-Preventing Plastic at Its Source: Recognizing that rivers act as plastic superhighways transporting waste from land to sea, The Ocean Clean-up has also introduced the Interceptor™ a solar-powered, barge-like device deployed in highly polluted rivers. These systems have been piloted in locations such as the Klang River in Malaysia and the Cengkareng Drain in Indonesia, where they have intercepted tons of floating debris before it reaches coastal waters (Lebreton et al., 2017).

Satellite and Drone Monitoring: Remote sensing technologies are now being leveraged to enhance the detection and mapping of plastic pollution. Satellite imagery, paired with artificial intelligence (AI) algorithms, can identify floating plastic accumulations across vast marine expanses. Similarly, drones are being used for coastal surveillance and targeted clean-up planning. These tools allow for real-time monitoring and more strategic allocation of clean-up resources (Agarwala, 2021).

Robotics and AI in Sorting: Robotics and AI are also revolutionizing post-collection sorting and recycling efforts. Smart marine robots, such as Jellyfishbot and WasteShark, are designed to patrol marinas and harbors, collecting floating debris with minimal environmental disruption. Meanwhile, AI-assisted sorting systems at recycling facilities help increase the purity and value of recovered plastic, enhancing the economic viability of closed-loop systems (Sarc et al., 2019).

Despite these advances, technological clean-up alone cannot solve the plastic crisis. These tools must be paired with strong prevention strategies, given that many clean-up solutions, especially those targeting microplastics or deep-sea debris, face limitations in reach, efficiency, and cost-effectiveness (Pellicer & Domingo, 2023).

Innovative filtration materials such as biodegradable nanocellulose membranes and bio-inspired filters mimicking mussel byssus are also under development, showing promise in removing a wide range of microplastic sizes and types. These technologies, though still emerging, represent the future of precision pollution control in both municipal and industrial contexts.

3.7 Conclusion

Plastic pollution has emerged as one of the most intractable environmental challenges of the twenty-first century, ubiquitous, persistent, and intricately woven into the fabric of modern industrial society. Its impacts extend across geographical, ecological, and socio-economic boundaries, from remote marine habitats and terrestrial ecosystems to public health, livelihoods, and global trade.

In summary, addressing the plastic pollution crisis is not solely an exercise in waste control; it is a transformative opportunity to reimagine our material culture, redefine value creation, and regenerate the ecological foundations on which all life depends. Through bold governance, interdisciplinary research, and inclusive innovation, we can forge a resilient and restorative future, one in which plastics are designed for circularity, communities thrive in harmony with their environment, and ecosystems are given the space and time to heal.

References

Agarwala, N. (2021). Managing marine environmental pollution using artificial intelligence. *Maritime Technology and Research, 3*(2), 120–136. https://doi.org/10.33175/mtr.2021.248053

Akhter, P., Nawaz, S., Shafiq, I., Nazir, A., Shafique, S., Jamil, F., Park, Y. K., & Hussain, M. (2023). Efficient visible light assisted photocatalysis using ZnO/TiO_2 nanocomposites. *Molecular Catalysis, 535*(November 2022), 112896. https://doi.org/10.1016/j.mcat.2022.112896

Akhtar, R., Sirwal, M. Y., Hussain, K., Dar, M. A., Shahnawaz, M., & Daochen, Z. (2022). Impact of plastic waste on the coral reefs: An overview. In M. Shahnawaz, M. K. Sangale, Z. Daochen, & A. B. Ade (Eds.), Impact of plastic waste on the marine biota (Chapter 13). Springer. https://doi.org/10.1007/978-981-16-5403-9_13

Alam, S. S., & Khan, A. H. (2024). Microwave-assisted pyrolysis for waste plastic recycling: A review on critical parameters, benefits, challenges, and scalability perspectives. *International Journal of Environmental Science and Technology, 21*(5), 5311–5330. https://doi.org/10.1007/s13762-023-05352-3

Alamgir, M., Islam, S. M. T., Saju, J. A., Khanal, R., & Ojha, B. (2023). Life cycle assessment of municipal plastic waste management in Kathmandu. *Journal of Material Cycles and Waste Management, 27*(1), 624–637. https://doi.org/10.1007/s10163-024-02098-z

Ali, S., Ahmed, W., Solangi, Y. A., Chaudhry, I. S., & Zarei, N. (2022). Strategic analysis of single-use plastic ban policy for environmental sustainability: The case of Pakistan. *Clean Technologies and Environmental Policy, 24*(3), 843–849. https://doi.org/10.1007/s10098-020-02011-w

Alpizar, F., Carlsson, F., Lanza, G., Carney, B., Daniels, R. C., Jaime, M., Ho, T., Nie, Z., Salazar, C., Tibesigwa, B., & Wahdera, S. (2020). A framework for selecting and designing policies to reduce marine plastic pollution in developing countries. *Environmental Science and Policy, 109*(April), 25–35. https://doi.org/10.1016/j.envsci.2020.04.007

Ammendolia, J., & Walker, T. R. (2024). Consistently inconsistent: The false promise of 'Sustainable' plastics. *Cambridge Prisms: Plastics, 2*, 2022–2025. https://doi.org/10.1017/plc.2024.9

Anon. (2017). Designing a plastic road in Netherlands.

Anon. (2022). *Global plastics outlook.*

Anon. (n.d.) Sector report circular economy senegal sector report circular economy Senegal.

Aragaw, T. A. (2025). Plastic waste management strategies toward zero waste: Status, perspectives and recommendations for Ethiopia *2018*(2024).

Arena, U. (2012). Process and technological aspects of municipal solid waste gasification. A review. *Waste Management, 32*(4), 625–639. https://doi.org/10.1016/j.wasman.2011.09.025

Asia, P. A. (2022). Challenges and efforts in Japan's plastic waste management—challenges.

Bakir, A., Rowland, S. J., & Thompson, R. C. (2014). Enhanced desorption of persistent organic pollutants from microplastics under simulated physiological conditions. *Environmental Pollution, 185*, 16–23. https://doi.org/10.1016/j.envpol.2013.10.007

Banu, N. (2020). Single-use plastic ban and its public health impacts: A narrative review. *Annals of SBV, 8*(1), 13–18. https://doi.org/10.5005/jp-journals-10085-8102

Barnes, D. K. A. (2002). Invasioni Della Vita Marina Su Detriti Di Plastica. *Natura, 416*(April), 808–809.

Bellou, N., Gambardella, C., Karantzalos, K., Monteiro, J. G., Canning-Clode, J., Kemna, S., Arrieta-Giron, C. A., & Lemmen, C. (2021). Global assessment of innovative solutions to tackle marine litter. *Nature Sustainability, 4*(6), 516–524. https://doi.org/10.1038/s41893-021-00726-2

Bhuyan, M. S., Jenzri, M., Pandit, D., Adikari, D., Alam, M. W., & Kunda, M. (2024). Microplastics occurrence in sea cucumbers and impacts on sea cucumbers & human health: A systematic review. *Science of the Total Environment, 951*(June), 175792. https://doi.org/10.1016/j.scitotenv.2024.175792

Boussellaa, W., Bradai, M. N., Mallat, H., Enajjar, S., Saidi, B., & Jribi, I. (2024). Ghost Gear in the Gulf of Gabès (Tunisia): An urgent need for a conservation code of conduct. *Sustainability (Switzerland), 16*(18). https://doi.org/10.3390/su16188003

Chae, Y., & An, Y. J. (2017). Effects of micro- and nanoplastics on aquatic ecosystems: Current research trends and perspectives. *Marine Pollution Bulletin, 124*(2), 624–632. https://doi.org/10.1016/j.marpolbul.2017.01.070

Chauhan, G., & Sevda, S. (2023). *Solid waste management: Chemical approaches, volume 1.*

Chertow, M., Reck, B. K., Wrzesniewski, A., & Calli, B. (2024). Outlook on the future role of robots and ai in material recovery facilities: Implications for U.S. recycling and the workforce. *Journal of Cleaner Production, 470*(December 2022), 143234. https://doi.org/10.1016/j.jclepro.2024.143234

Coyle, R., Hardiman, G., & O' Driscoll, K. (2020). Microplastics in the marine environment: A review of their sources, distribution processes, uptake and exchange in ecosystems. *Case Studies in Chemical and Environmental Engineering, 2*(May). https://doi.org/10.1016/j.cscee.2020.100010

Danielli, F., & Sousa, B. D. (2025). The global plastics treaty: Understanding the present to guide the future.

Danso, D., Chow, J., & Streita, W. R. (2019). Plastics: Environmental and biotechnological perspectives on microbial degradation. *Applied and Environmental Microbiology, 85*(19). https://doi.org/10.1128/AEM.01095-19

Do, H. L., & Armstrong, C. W. (2023). Ghost fishing gear and their effect on ecosystem services—identification and knowledge gaps. *Marine Policy, 150*(June 2022), 105528. https://doi.org/10.1016/j.marpol.2023.105528

Ellen Macarthur Foundation; World Economic Forum; McKinsey & Co. (2016). The new plastic economy: Rethinking the future of plastics. *Ellen Macarthur Foundation* (January), 1–120.

van Emmerik, T., Vriend, P., & Peereboom, E. C. (2022). Roadmap for long-term macroplastic monitoring in rivers. *Frontiers in Environmental Science, 9*(February), 1–8. https://doi.org/10.3389/fenvs.2021.802245

Erdle, L. M., Parto, D. N., Sweetnam, D., & Rochman, C. M. (2021). Washing machine filters reduce microfiber emissions: Evidence from a community-scale pilot in parry sound, Ontario. *Frontiers in Marine Science, 8*(November), 1–9. https://doi.org/10.3389/fmars.2021.777865

Flygansvær, B., & Dahlstrom, R. (2024). Enhancing circular supply chains via ecological packaging: An empirical investigation of an extended producer responsibility network. *Journal of Cleaner Production, 468*(May), Article 142948. https://doi.org/10.1016/j.jclepro.2024.142948

Forrest, A., Giacovazzi, L., Dunlop, S., Reisser, J., Tickler, D., Jamieson, A., & Meeuwig, J. J. (2019). Eliminating plastic pollution: How a voluntary contribution from industry will drive the circular plastics economy. *Frontiers in Marine Science, 6*(SEP), 1–11. https://doi.org/10.3389/fmars.2019.00627

Fotopoulou, K. N., & Karapanagioti, H. K. (2019). Degradation of various plastics in the environment. *Handbook of Environmental Chemistry, 78*, 71–92. https://doi.org/10.1007/698_2017_11

Frantzi, S., Brouwer, R., Watkins, E., van Beukering, P., Cunha, M. C., Dijkstra, H., Duijndam, S., Jaziri, H., Okoli, I. C., Pantzar, M., Cotera, I. R., Rehdanz, K., Seidel, K., & Triantaphyllidis, G. (2021). Adoption and diffusion of marine litter clean-up technologies across European seas: Legal, institutional and financial drivers and barriers. *Marine Pollution Bulletin, 170*(February 2021). https://doi.org/10.1016/j.marpolbul.2021.112611

Galgani, L., Beiras, R., Galgani, F., Panti, C., & Borja, A. (2019). Editorial: 'Impacts of Marine Litter.' *Frontiers in Marine Science, 6*(APR), 4–7. https://doi.org/10.3389/fmars.2019.00208

García-Marín, L. M., & Rentería, M. E. (2024). Fighting plastic pollution with a circular economy roadmap and strategy: Addressed to the United Nations environment programme. *Journal of Science Policy & Governance, 24*(01). https://doi.org/10.38126/jspg240107

Geyer, R., Jambeck, J. R., & Law, K. L. (2017). Production, use, and fate of all plastics ever made. *Science Advances, 3*(7), 25–29. https://doi.org/10.1126/sciadv.1700782

Gkanasos, A., Tsiaras, K., Triantaphyllidis, G., Panagopoulos, A., Pantazakos, G., Owens, T., Karametsis, C., Pollani, A., Nikoli, E., Katsafados, N., & Triantafyllou, G. (2021). Stopping macroplastic and microplastic pollution at source by installing novel technologies in river estuaries and waste water treatment plants: The CLAIM project. *Frontiers in Marine Science, 8*(December). https://doi.org/10.3389/fmars.2021.738876

Government of the UK. (2023). Plastic packaging tax—GOV.UK. *Gov.Uk* (May).

Hermabessiere, L., Dehaut, A., Paul-Pont, I., Lacroix, C., Jezequel, R., Soudant, P., & Duflos, G. (2017). Occurrence and effects of plastic additives on marine environments and organisms: A review. *Chemosphere, 182*, 781–793. https://doi.org/10.1016/j.chemosphere.2017.05.096

Hidayat, D. S. W. (2025). Efforts to handle waste management through clean village program planning in improving environmental quality (Case Study Of Batang Kuis District) *2*(1), 214–230.

Hopewell, J., Dvorak, R., & Kosior, E. (2009). Plastics recycling: Challenges and opportunities. *Philosophical Transactions of the Royal Society B: Biological Sciences, 364*(1526), 2115–2126. https://doi.org/10.1098/rstb.2008.0311

Iyare, P. U., Ouki, S. K., & Bond, T. (2020). Microplastics removal in wastewater treatment plants: A critical review. *Environmental Science: Water Research and Technology, 6*(10), 2664–2675. https://doi.org/10.1039/d0ew00397b

Jambeck, J., Geyer, R., Wilcox, C., Siegler, T. R., Perryman, M., Andrady, A., Narayan, R., & Law, K. L. (2015). The Ocean: The Ocean. *Marine Pollution, 347*(6223), 768.

Kaminsky, W. (2021). Chemical recycling of plastics by fluidized bed pyrolysis. *Fuel Communications, 8*(May), Article 100023. https://doi.org/10.1016/j.jfueco.2021.100023

Kasznik, D., & Łapniewska, Z. (2023). The end of plastic? The EU's directive on single-use plastics and its implementation in Poland. *Environmental Science and Policy, 145*(March), 151–163. https://doi.org/10.1016/j.envsci.2023.04.005

Kaza, S., Yao, L., Bhada-Tata, P., & Van Woerden, F. (2018). What a waste 2.0 introduction—"Snapshot of Solid Waste Management to 2050." Overview booklet. *Urban Development Series*, 1–38.

Kibria, M. G., Masuk, N. I., Safayet, R., et al. (2023). Plastic waste: Challenges and opportunities to mitigate pollution and effective management. *International Journal of Environmental Research, 17*(20). https://doi.org/10.1007/s41742-023-00507-z

Kumar, R., Verma, A., Shome, A., Sinha, R., Sinha, S., Jha, P. K., Kumar, R., Kumar, P., Shubham, S. D., Sharma, P., & Vara Prasad, P. V. (2021). Impacts of plastic pollution on ecosystem services, sustainable development goals, and need to focus on circular economy and policy interventions. *Sustainability (Switzerland), 13*(17), 1–40. https://doi.org/10.3390/su13179963

Lamb, J. B., Willis, B. L., Fiorenza, E. A., Couch, C. S., Howard, R., Rader, D. N., True, J. D., Kelly, L. A., Ahmad, A., Jompa, J., & Drew Harvell, C. (2018). Plastic waste associated with disease on coral reefs. *Science, 359*(6374), 460–462. https://doi.org/10.1126/science.aar3320

Leadership, C. F. O. (n.d.). CFO leadership network responding to the worked example: Unilever.

Leal Filho, W., Barbir, J., May, J., May, M., Swart, J., Yang, P., Dinis, M. A. P., Aina, Y., Bettencourt, S., Charvet, P., & Azadi, H. (2025). Towards more sustainable oceans: A review of the pressing challenges posed by marine plastic litter. *Waste Management & Research*. https://doi.org/10.1177/0734242X251313927

Lebreton, L. C. M., Van Der Zwet, J., Damsteeg, J. W., Slat, B., Andrady, A., & Reisser, J. (2017). River plastic emissions to the world's oceans. *Nature Communications, 8*, 1–10. https://doi.org/10.1038/ncomms15611

Lee, B. X., Kjaerulf, F., Turner, S., Cohen, L., Donnelly, P. D., Muggah, R., Davis, R., Realini, A., Kieselbach, B., MacGregor, L. S., Waller, I., Gordon, R., Moloney-Kitts, M., Lee, G., & Gilligan, J. (2016). Transforming our world: Implementing the 2030 agenda through sustainable development goal indicators. *Journal of Public Health Policy, 37*(1), S13-31. https://doi.org/10.1057/s41271-016-0002-7

Lee, Q. Y., & Li, H. (2021). Photocatalytic degradation of plastic waste: A mini review. *Micromachines, 12*(8). https://doi.org/10.3390/mi12080907

Lee, S., & Na, S. (2010). E-waste recycling systems and sound circulative economies in east Asia: A comparative analysis of systems in Japan, South Korea, China and Taiwan. *Sustainability, 2*(6), 1632–1644. https://doi.org/10.3390/su2061632

Li, Y., Zhang, H., & Tang, C. (2020). A review of possible pathways of marine microplastics transport in the ocean. *Anthropocene Coasts, 3*(1), 6–13. https://doi.org/10.1139/anc-2018-0030

Litaudon, V., & Chen, Y. (2023). Transforming existing local ecosystems to circular economy ecosystems for plastics: The case of the PlastiCity ecosystem. *Journal of Industrial Ecology, 27*(5), 1406–1419. https://doi.org/10.1111/jiec.13433

Lithner, D., Larsson, A., & Dave, G. (2011). Environmental and health hazard ranking and assessment of plastic polymers based on chemical composition. *Science of the Total Environment, 409*(18), 3309–3324. https://doi.org/10.1016/j.scitotenv.2011.04.038

Lopez, G., Artetxe, M., Amutio, M., Alvarez, J., Bilbao, J., & Olazar, M. (2018). Recent advances in the gasification of waste plastics. A critical overview. *Renewable and Sustainable Energy Reviews, 82*(September 2017), 576–596. https://doi.org/10.1016/j.rser.2017.09.032

Luo, Z., Zhou, X., Su, Y., Wang, H., Yu, R., Zhou, S., Xu, E. G., & Xing, B. (2021). Environmental occurrence, fate, impact, and potential solution of tire microplastics: Similarities and differences with tire wear particles. *Science of the Total Environment, 795*. https://doi.org/10.1016/j.scitotenv.2021.148902

Makarichi, L., Jutidamrongphan, W., & Okpara, K. (2023). Choosing among plastic waste management options: Lessons from Zimbabwe's plastic waste flows. *The Lancent Pschch, 11*(maret), 133–143. https://doi.org/10.1007/s10163-024-02132-0

Marshall, M. (2018). *From pollution to solution* (Vol. 237).

Martin, C., Almahasheer, H., & Duarte, C. M. (2019). Mangrove forests as traps for marine litter. *Environmental Pollution, 247*, 499–508. https://doi.org/10.1016/j.envpol.2019.01.067

Martín, J., Santos, J. L., Aparicio, I., & Alonso, E. (2022). Microplastics and associated emerging contaminants in the environment: Analysis, sorption mechanisms and effects of co-exposure. *Trends in Environmental Analytical Chemistry, 35*(June), Article e00170. https://doi.org/10.1016/j.teac.2022.e00170

Matavos-Aramyan, S. (2024). Addressing the microplastic crisis: A multifaceted approach to removal and regulation. *Environmental Advances, 17*(June 2023), 100579. https://doi.org/10.1016/j.envadv.2024.100579

Miksch, L., Köck, M., Gutow, L., & Saborowski, R. (2022). Bioplastics in the sea: Rapid in-vitro evaluation of degradability and persistence at natural temperatures. *Frontiers in Marine Science, 9*(June), 1–12. https://doi.org/10.3389/fmars.2022.920293

Nanda, A., Panda, S., & Panigrahi, S. K. (2025). Production of durable conventional concrete using recycled HDPE and PET plastic coarse aggregate. *Innovative Infrastructure Solutions, 10*(3). https://doi.org/10.1007/s41062-025-01886-2

Nduwimana, J., Ndikumana, T., & Luis, P. (2026). Plastic waste management policies in the east African community (EAC) region: Challenges and prospects. *2015*(1), 112–127.

Nikiema, J., & Asiedu, Z. (2022). A review of the cost and effectiveness of solutions to address plastic pollution. *Environmental Science and Pollution Research, 29*(17), 24547–24573. https://doi.org/10.1007/s11356-021-18038-5

Omeyer, L. C. M., Duncan, E. M., Abreo, N. A. S., Acebes, J. M. V., AngSinco-Jimenez, L. A., Anuar, S. T., Aragones, L. V., Araujo, G., Carrasco, L. R., Chua, M. A. H., Cordova, M. R., Dewanti, L. P., Espiritu, E. Q., Garay, J. B., Germanov, E. S., Getliff, J., Horcajo-Berna, E., Ibrahim, Y. S., Jaafar, Z., …, Godley, B. J. (2023). Interactions between marine megafauna and plastic pollution in southeast Asia. *Science of the Total Environment, 874*(November 2022), 162502. https://doi.org/10.1016/j.scitotenv.2023.162502

Paul, W. M. K., & Mironga, J. M. (2020). Effectiveness of the implementation of plastic bags ban: Empirical evidence from Kenya. *IOSR Journal of Environmental Science, 14*(6), 53–61. https://doi.org/10.9790/2402-1406025361

Pellicer, I., & Domingo, F. C. (2023). Microplastic pollution in the coast of Tarragona, Spain: A western mediterranean study to cite this version: HAL Id: Hal-04174194.

Prata, J. C., da Costa, J. P., Lopes, I., Duarte, A. C., & Rocha-Santos, T. (2020). Environmental exposure to microplastics: An overview on possible human health effects. *Science of the Total Environment, 702*, Article 134455. https://doi.org/10.1016/j.scitotenv.2019.134455

Ragaert, K., Delva, L., & Van Geem, K. (2017). Mechanical and chemical recycling of solid plastic waste. *Waste Management, 69*, 24–58. https://doi.org/10.1016/j.wasman.2017.07.044

Rangel-Buitrago, N., Neal, W. J., & Galgani, F. (2023). Plastics in the anthropocene: A multifaceted approach to marine pollution management. *Marine Pollution Bulletin, 194*(PA), 115359. https://doi.org/10.1016/j.marpolbul.2023.115359

Russell, J. R., Huang, J., Anand, P., Kucera, K., Sandoval, A. G., Dantzler, K. W., Hickman, D. S., Jee, J., Kimovec, F. M., Koppstein, D., Marks, D. H., Mittermiller, P. A., Núñez, S. J., Santiago, M., Townes, M. A., Vishnevetsky, M., Williams, N. E., Vargas, M. P. N., Boulanger, L. A., … Strobel, S. A. (2011). Biodegradation of polyester polyurethane by endophytic fungi. *Applied and Environmental Microbiology, 77*(17), 6076–6084. https://doi.org/10.1128/AEM.00521-11

SAPEA (2019). SAPEA, science advice for policy by European academies. A scientific perspective on microplastics in nature and society.

Sarc, R., Curtis, A., Kandlbauer, L., Khodier, K., Lorber, K. E., & Pomberger, R. (2019). Digitalisation and intelligent robotics in value chain of circular economy oriented waste management—a review. *Waste Management, 95*, 476–492. https://doi.org/10.1016/j.wasman.2019.06.035

Setälä, O., Fleming-Lehtinen, V., & Lehtiniemi, M. (2014). Ingestion and transfer of microplastics in the planktonic food web. *Environmental Pollution, 185*, 77–83. https://doi.org/10.1016/j.envpol.2013.10.013

Singh, N., Hui, D., Rupinder Singh, I. P. S., Ahuja, L. F., & Fraternali, F. (2017). Recycling of plastic solid waste: A state of art review and future applications. *Composites Part B: Engineering, 115*, 409–422. https://doi.org/10.1016/j.compositesb.2016.09.013

Sinha, R. K., Kumar, R., Phartyal, S. S., & Sharma, P. (2024). Interventions of citizen science for mitigation and management of plastic pollution: Understanding sustainable development goals, policies, and regulations. *Science of the Total Environment, 955*(July), Article 176621. https://doi.org/10.1016/j.scitotenv.2024.176621

Slat, B. (2023). Where I work boyan slat. *Nature.*

Sun, J., Dai, X., Wang, Q., van Loosdrecht, M. C. M., & Ni, B. J. (2019). Microplastics in wastewater treatment plants: Detection, occurrence and removal. *Water Research, 152*, 21–37. https://doi.org/10.1016/j.watres.2018.12.050

Term, S. (2022). Roadmap to plastic-free UP diliman (2022), 1–7.

Tewary, A., & Upadhyay, P. (2024). Proposing a 6R framework promoting circular strategies for platform organizations. *Digital Policy, Regulation and Governance, Advance Online Publication.* https://doi.org/10.1108/DPRG-11-2023-0169.

Thompson, R. C., Moore, C. J., Votn Saal, F. S., & Swan, S. H. (2009). Plastics, the environment and human health: Current consensus and future trends. *Philosophical Transactions of the Royal Society B. Biological Sciences, 364*(1526), 2153–2166. https://doi.org/10.1098/rstb.2009.0053

Tournier, V., Topham, C. M., Gilles, A., David, B., Folgoas, C., Moya-Leclair, E., Kamionka, E., Desrousseaux, M. L., Texier, H., Gavalda, S., Cot, M., Guémard, E., Dalibey, M., Nomme, J., Cioci, G., Barbe, S., Chateau, M., André, I., Duquesne, S., & Marty, A. (2020). An engineered PET depolymerase to break down and recycle plastic bottles. *Nature, 580*(7802), 216–219. https://doi.org/10.1038/s41586-020-2149-4

United Nations Environment Programme. (2024, February 19). Annual report 2023. https://www.unep.org/resources/annual-report-2023.

Vaquer-Sunyer, R., & Duarte, C. M. (2008). Thresholds of hypoxia for marine biodiversity. *Proceedings of the National Academy of Sciences of the United States of America, 105*(40), 15452–15457. https://doi.org/10.1073/pnas.0803833105

Vergara, D., de la Hoz-M, J., Ariza-Echeverri, E. A., Fernández-Arias, P., & Antón-Sancho, Á. (2025). Evaluating solutions to marine plastic pollution. *Environments—MDPI, 12*(3). https://doi.org/10.3390/environments12030086

Wali, S. U. (2021). The need for a multi-pollutant approach to model the movement of pollutants in surface-water: A review of status and future challenges. *International Journal of Agriculture and Animal Production,* (August), 26–58. https://doi.org/10.55529/ijaap.11.26.58

Wei, R., & Zimmermann, W. (2017). Microbial enzymes for the recycling of recalcitrant petroleum-based plastics: How far are we? *Microbial Biotechnology, 10*(6), 1308–1322. https://doi.org/10.1111/1751-7915.12710

Widiastutie, S., Maarif, D., Saraswati, D. P., Sianipar, I. M. J., Phan, T. T. T., & Nguyen, V. V. (2025). Residents' adaptive marine plastic litter management in Galang Island, Indonesia: An importance-performance analysis. *Marine Policy, 172*(October 2024). https://doi.org/10.1016/j.marpol.2024.106508

Wilcox, C., Heathcote, G., Goldberg, J., Gunn, R., Peel, D., & Hardesty, B. D. (2015). Understanding the sources and effects of abandoned, lost, and discarded fishing gear on marine turtles in northern Australia. *Conservation Biology, 29*(1), 198–206. https://doi.org/10.1111/cobi.12355

Wilson, D. C., Velis, C., & Cheeseman, C. (2006). Role of informal sector recycling in waste management in developing countries. *Habitat International, 30*(4), 797–808. https://doi.org/10.1016/j.habitatint.2005.09.005

Wright, S. L., & Kelly, F. J. (2017). Plastic and human health: A micro issue? *Environmental Science and Technology, 51*(12), 6634–6647. https://doi.org/10.1021/acs.est.7b00423

Yang, Y., Yang, J., & Jiang, L. (2016). Comment on "a Bacterium That Degrades and Assimilates Poly(Ethylene Terephthalate)." *Science, 353*(6301), 759. https://doi.org/10.1126/science.aaf8305

Yoshida, S., Hiraga, K., Takehana, T., Taniguchi, I., Yamaji, H., Maeda, Y., Toyohara, K., Miyamoto, K., Kimura, Y., & Oda, K. (2016). Una Bacteria Que Degrada y Asimila Poli (Tereftalato de Etileno). *Science, 353*(6301), 759.

Zafar, M. A., & Jacob, M. V. (2024). Instant upcycling of microplastics into graphene and its environmental application. *Small Science, 2400176*, 1–8. https://doi.org/10.1002/smsc.202400176

Zhou, Y., Kumar, M., Sarsaiya, S., Sirohi, R., Awasthi, S. K., Sindhu, R., Binod, P., Pandey, A., Bolan, N. S., Zhang, Z., Singh, L., Kumar, S., & Awasthi, M. K. (2022). Challenges and opportunities in bioremediation of micro-nano plastics: A review. *Science of the Total Environment, 802*, Article 149823. https://doi.org/10.1016/j.scitotenv.2021.149823

4 Future Prospects for a Sustainable Environment

4.1 Introduction

The environmental challenges posed by plastics have been thoroughly explored in previous chapters, highlighting how their widespread use across packaging, healthcare, electronics, transportation, and construction has led to unprecedented levels of pollution. Plastics' inherent durability, while beneficial for industrial applications, has resulted in their persistent accumulation across ecosystems. Traditional waste management approaches, including landfilling, incineration, and mechanical recycling, have proven insufficient to handle the massive volumes generated (Lange, 2021; Wang et al., 2022). Issues such as microplastic contamination, marine debris, greenhouse gas emissions from degradation, and the leaching of toxic additives into soil and water systems have compounded the crisis. Despite global awareness efforts and initial policy responses, the recycling rate of plastics remains alarmingly low, and production continues to grow at an unsustainable pace.

Given the limitations of current systems, there is an urgent need to shift away from a linear "take-make-dispose" approach and toward a more circular and sustainable paradigm. This shift needs creative, scientifically grounded solutions that can clean up current waste, stop future plastic pollution, and encourage the use of regenerative materials (Clark & Shaver, 2024). Encouragingly, there is growing interest among academia, corporations, governments, and communities in exploring long-term, integrated initiatives that incorporate technology, policy, and education. Future prospects involve a combination of technological innovation, policy evolution, and behavioral transformation aimed at building a circular and regenerative plastic economy. Advanced recycling technologies, plastic pollution monitoring systems, biological remediation strategies, and the adoption of nature-based and circular economy models represent promising directions (Alaghemandi, 2024). This chapter examines these emerging approaches, beginning with methods

N. T. Hatvate et al., *Plastic Waste Management*, Synthesis Lectures on Sustainable Development, https://doi.org/10.1007/978-3-031-96660-6_4

to measure plastic pollution, followed by innovations in recycling and remediation technologies, circular economy frameworks, and forward-looking policy recommendations. By integrating scientific advancement with governance and public engagement, a more sustainable and resilient environmental future becomes achievable.

4.2 Methods for Measuring Plastic Pollution

Accurate measurement and monitoring are the first steps toward effective environmental management of plastic waste. As plastic pollution becomes increasingly complex, particularly with the rise of microplastics and nanoplastics, the ability to detect, measure, and characterize plastic contaminants in different environmental settings has become a pillar of sustainability research. Understanding the amount, sources, and pathways of plastic waste is critical for developing mitigation strategies, shaping policies, and safeguarding both ecosystems and public health.

4.2.1 The Challenge of Microplastic Detection

Microplastics (MPs), which are plastic particles smaller than 5 mm in size, are among the most insidious pollutants due to their prevalence, persistence, and biological effects (Matavos-Aramyan, 2024; Sarkar et al., 2023). They originate from two main sources:

- Primary microplastics are produced at microscopic scales for use in cosmetics, cleaning goods, and industrial activities.
- Secondary microplastics are created when bigger plastic waste fragments due to physical, chemical, or biological degradation.

MPs can accumulate in sediments, float on water surfaces, or embed in biological tissues in aquatic environments, especially in aquaculture systems and marine ecosystems. According to studies, they can be found in human blood and lungs as well as in seafood, table salt, and drinking water (Barceló et al., 2023; Sarkar et al., 2023). Their capacity to absorb and transmit harmful substances, like heavy metals and PCBs, increases the dangers to human health and the environment.

4.2.2 Sampling and Collection Methods

In microplastic analysis, sampling is a crucial first step, and different approaches are used based on the environmental matrix (water, sediment, or biota). Important strategies consist of Volume-reduced sampling, bulk sampling, and selective samplings.

4.2.2.1 Volume-Reduced Sampling

This technique includes pre-filtering huge amounts of water on-site with sieves or nets to allow for particle concentration prior to laboratory analysis. Neuston or manta trawls are frequently employed for sampling surface water, and their mesh sizes are approximately 333 μm.

4.2.2.2 Bulk Sampling

This technique involves taking samples of whole water or sediment without prior filtration. These are then processed in the lab, which allows for greater control but requires deeper analysis.

4.2.2.3 Selective Sampling

Collects visible plastic particles straight from the environment. Although simple, it may not capture smaller or submerged MPs.

Mesh size is an important factor in determining the detection threshold. Smaller mesh sizes improve fine particle recovery while increasing the risk of blockage from organic materials. While tissue dissection is usually required for biota sample in order to extract ingested MPs, stainless steel scoops or corers are employed for sediments (Prata et al., 2019; Sharma et al., 2024; Tasseron et al., 2024).

4.2.3 Separation and Purification Techniques

Samples must be carefully separated from organic and inorganic materials after collection. The following methods are used for separation and purifications.

4.2.3.1 Density Separation

Based on the fact that water or sediments have a higher density than most polymers. Denser solutions, such as zinc chloride ($ZnCl_2$) or sodium iodide (NaI), can better separate heavier plastics like PET and PVC than saturated salt solutions (e.g., NaCl). Despite their efficiency, these reagents need to be handled carefully because they are expensive and perhaps dangerous.

4.2.3.2 Digestion

Removes biological material that may obstruct analysis.

Hydrogen peroxide (H_2O_2) is a commonly used and effective method for oxidative digestion of organic materials. Acidic or alkaline digestion (e.g., nitric acid, potassium hydroxide) may be more aggressive, but it can harm plastic polymers. Although it is costly and time-consuming, enzymatic digestion provides a more environmentally friendly and plastic-safe substitute.

Filtering and sieving are frequently employed to separate MPs according to size. Due to their compatibility with spectroscopic analysis, membrane filters, such as glass fiber and cellulose nitrate, are preferred. Consistent recovery is guaranteed by proper filtration, although particles smaller than 20 μm might still be lost (Parga Martínez et al., 2023; Prata et al., 2019; Schrank et al., 2022).

4.2.3.3 Analytical Identification and Quantification

Robust analytical procedures are necessary to identify plastic types and confirm that particles are, in fact, synthetic polymers:

- Visual microscopy: The particles are visually inspected and arranged according to their size, shape, and color using stereomicroscopes or scanning electron microscopy (SEM) (Kotar et al., 2022; Mariano et al., 2021). However, organic debris can be mistaken for plastics using only visual approaches.
- Infrared Fourier Transform Spectroscopy (FTIR): This approach uses infrared absorption spectra to identify polymers. Because of its dependability, it is frequently employed; nonetheless, it is sensitive to sample contaminants and has limitations in identifying particles smaller than 20 μm (Campanale et al., 2023).
- Raman Spectroscopy: Strengthens FTIR by allowing higher spatial resolution and better detection of smaller MPs (down to 1 μm), even in wet samples. However, the problem of fluorescent interference still exists (Jung et al., 2024).
- Pyrolysis–Gas Chromatography/Mass Spectrometry (Py-GC/MS): It is a destructive approach in which MPs are thermally broken down into monomers and identified using their chemical structure. It makes it possible to identify and measure several plastic types at once, particularly in complex or highly degraded samples (Kwon et al., 2025).

4.3 Plastic Recycling Technologies: Research and Development

Plastic waste management has progressed from basic disposal techniques to advanced recycling technologies that support a circular economy while simultaneously attempting to reduce environmental impact. Traditional approaches, such as mechanical recycling and landfilling, have shown shortcomings in terms of sustainability, efficiency, and scale (Achilias, 2025). As a result, research and development (R&D) activities are currently concentrated on innovative plastic recycling technologies that provide superior environmental outcomes, increased material recovery rates, and increased energy efficiency.

4.3.1 Mechanical Recycling

Mechanical recycling, or physical reprocessing, is currently the most widely utilized technique worldwide. Plastics are sorted, washed, shredded, melted, and re-granulated for reuse. In clean, single-stream waste, this method is especially effective for thermoplastics such as polyethylene (PE), polypropylene (PP), and polyethylene terephthalate (PET). However, handling polluted, colored, or mixed plastics significantly reduces its efficacy, resulting in lower-quality recycled or downcycled goods (Alaghemandi, 2024; Umdagas et al., 2025).

Recent R&D initiatives are addressing these inefficiencies through:

- AI-powered sorting devices that improve accuracy in mixed garbage streams by automatically classifying plastic kinds using machine learning algorithms, hyperspectral imaging, and near-infrared spectroscopy (Olawade et al., 2024).
- Tracer-based sorting (TBS), which improves material separation accuracy during post-consumer sorting by labeling particular polymer types with detectable markers (Gasde et al., 2021).
- In blended recycling processes, compatibilizer technologies involve the addition of chemical agents that increase the miscibility of incompatible polymers. These chemicals give recycled blends their mechanical strength and structural integrity again (Dorigato, 2021).
- Before remolding, decontamination techniques like enzyme washing, ultrasonic cleaning, and supercritical CO_2 treatment are being developed to get rid of harmful compounds, labels, and dyes from post-consumer plastic (Alaghemandi, 2024).

While mechanical recycling remains cost-effective and readily available, the most significant difficulty is the degradation of polymer chains during thermal processing. This results in decreased mechanical properties in recycled polymers, limiting their usage in high-end applications.

4.3.2 Chemical Recycling

Chemical recycling, often known as feedstock recycling, is a transformative approach that breaks down polymers into their constituent monomers or simpler hydrocarbons. These recovered building blocks can be recycled to create virgin-quality polymers, fuels, or chemical feedstocks, thereby closing the material loop.

The following are the primary types of chemical recycling:

- Pyrolysis: A thermal deterioration process carried out in the absence of oxygen. It produces liquid oils, gasses, and solid residues from mixed or contaminated plastics.

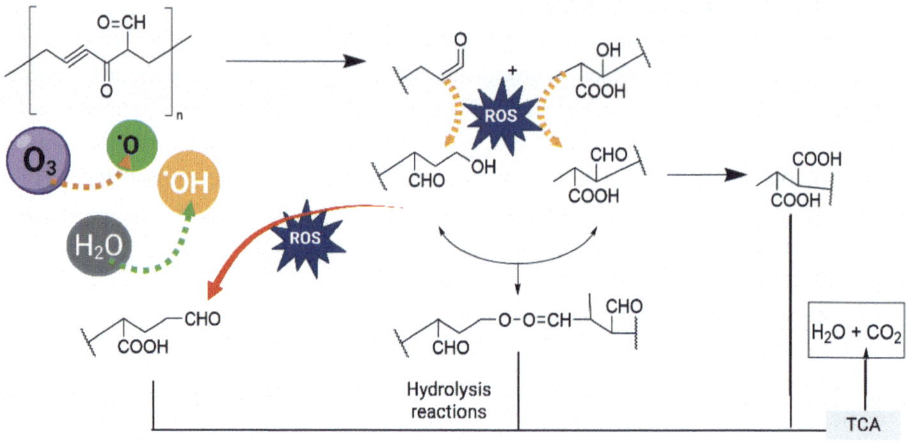

Fig. 4.1 Thermal degradation of plastic (*Created by using Biorender.com*)

Pyrolysis works very well for polyolefins like PE and PP. The resultant product can be processed into fuels for transportation or utilized as feedstock in the petrochemical sector. The process is energy-intensive, though, and there are still challenges to commercialization (Mong et al., 2022). The mechanism of thermal degradation is shown in Fig. 4.1.

- Gasification: This process converts plastics into synthesis gas (syngas), a blend of carbon monoxide and hydrogen while operating in an oxygen-limited atmosphere. Through catalytic synthesis, syngas can be transformed into new polymers, methanol, or ammonia or used to create energy. Gasification provides more energy recovery and lower emissions than incineration (Saebea et al., 2020).
- Depolymerization/Chemolysis: This includes hydrolysis, methanolysis, and glycolysis, which are particularly effective for polyester-based polymers like PET and polyamides (PA). These processes break down polymer chains into purifiable and re-polymerizable monomers, such as ethylene glycol and terephthalic acid (McNeeley & Liu, 2024).
- Solvent-based purification: New techniques such as the Solvent-Targeted Recovery and Precipitation (STRAP) process dissolve composite or multilayer plastics in specific solvents, enabling the recovery of pure polymers without structural degradation and the elimination of additives (Ikegwu et al., 2025).

Chemical recycling systems, despite their potential, are still under development. High operating costs, hazardous wastes, and the requirement for strong regulatory frameworks to guarantee both economic viability and environmental safety are challenges.

4.3.3 Biological Recycling

Biological degradation, which uses enzymes or microbes, is one of the most inventive recycling technologies. Although polymers like PET and PLA have the potential to biodegrade under certain circumstances, their breakdown rates in natural environments are extremely low. Scientists have discovered and modified microbial strains that can break down plastics by secreting enzymes. The mechanism of microbial degradation is shown in Fig. 4.3.

- PETase and MHETase: Ideonella sakaiensis and other bacteria have been found to produce enzymes that can depolymerize PET into monomers such as terephthalic acid and mono-(2-hydroxyethyl) terephthalate (MHET). These enzymes function in mild conditions to generate reusable monomers (Sevilla et al., 2023; Singh Jadaun et al., 2022).
- Fungal and bacterial degradation: Several fungal species (Aspergillus, Penicillium) and bacterial genera (Pseudomonas, Bacillus) may break down polyurethane (PU), polyethylene, and polystyrene. Degradation rates are slow, though, and most bacteria require special pre-treatment before they can completely mineralize polymers into CO_2 and water (Bhavsar et al., 2023; Park et al., 2023).
- Enzyme engineering and bioreactor development: Advanced genetic modification and protein engineering are employed to increase the thermostability, selectivity, and effectiveness of plastic-degrading enzymes. The commercial viability of these techniques is being tested in pilot-scale bioreactors (Joho et al., 2024).

Biological recycling, however, is still in its early stages and provides a low-energy, environmentally benign substitute for thermal or chemical treatments, especially for polymers that are challenging to recycle mechanically.

Table 4.1 outlines essential technologies and approaches, emphasizing their benefits and contributions to both waste management and economic development. Innovations such as AI-enabled sorting systems, advanced pre-treatment techniques, enzyme-assisted recycling, and microbial degradation significantly improve the effectiveness and quality of plastic recycling. These solutions help mitigate environmental pollution while fostering environmentally responsible economic growth. By overcoming key limitations of conventional recycling methods, these advancements support the transition toward a more circular and sustainable plastics economy.

Table 4.1 Advantages and impacts of key innovations in plastic waste recycling and circular economy

Innovation/technique	Advantages	Impact on recycling and circular economy
AI-driven sorting systems	High accuracy, reduced labor, handles complex streams	Improves purity of recyclates, enhances sorting efficiency, boosts recycling rates
Tracer-Based Sorting (TBS)	Fluorescent markers for precise polymer identification	Facilitates high-quality recycling, critical for multilayer and mixed plastics
Solvent-based pre-treatment	Removes contaminants while preserving polymer integrity	Enhances quality and usability of recycled plastic, reduces post-processing cost
Advanced extrusion equipment	Energy-efficient, precise temperature control, degassing systems	Minimizes polymer degradation, lowers energy use, improves product quality
Compatibilization strategies	Blends incompatible plastics, improves mechanical properties	Expands range of recyclable materials, enhances economic feasibility of recycling mixed plastics
Catalytic chemical recycling	Converts plastics to high-value monomers and chemicals	Reduces reliance on virgin inputs, adds economic value, supports circular feedstocks
Modular chemical recycling units	Scalable, flexible deployment	Decentralizes recycling, lowers logistics cost, supports local circular economies
Enzyme-based recycling	Operates under mild conditions, specific to polymers like PET	Enables closed-loop recycling, environmentally friendly alternative to thermal processing
Microbial degradation	Naturally degrades plastics, low energy requirement	Restores polluted sites, viable for bioremediation, complements existing recycling infrastructure

4.4 Remediation of Plastic Waste

Comprehensive remediation techniques are required to address the worldwide ecological challenge posed by the fast accumulation of plastic garbage in terrestrial, marine, and atmospheric habitats. For decades, plastic debris remains in ecosystems, breaking down into micro- and nanoplastics that are dangerous to aquatic life, plant systems, soil health, and human health. As a result, a broad range of physical, chemical, and biological techniques have been incorporated into remediation technologies to help break down, remove, or reuse plastic trash.

4.4.1 Physical Remediation Techniques

The first line of defense in plastic remediation is physical method. They mostly involve gathering, sorting, and reducing the size of plastic waste in order to get it prepared for additional processing. Two important advances in the extensive removal of aquatic plastic are the Great Bubble Barrier and ocean clean-up interceptors. One Dutch implementation reported an 86% plastic removal efficiency from canal waters using these devices, which use passive collecting techniques and underwater barriers to capture floating waste without endangering aquatic life (Leone et al., 2023).

However, because of their small size and widespread distribution, physical methods frequently fail to effectively treat micro- and nanoplastics. Filters, sedimentation, and flotation techniques are being improved to improve the effectiveness of removing these tiny particles from wastewater and natural bodies.

4.4.2 Chemical and Advanced Oxidation Processes (AOPs)

Chemical degradation converts polymers into less hazardous or reusable materials. Chemical degradation converts polymers into less hazardous or reusable materials. High molecular weight polymers are broken down into smaller molecules under heat, such as pyrolysis, which produces fuels and monomers (McNeeley & Liu, 2024). Despite its efficiency, it usually consumes a lot of energy and works best in regulated environments.

The potential of advanced oxidation techniques (AOPs) to mineralize plastic pollutants in situ is drawing attention. Examples of these methods include photocatalysis and electrochemical oxidation. Reactive oxygen species (ROS), like hydroxyl radicals (•OH), are produced by these processes and readily attack polymer strands. Photocatalysis with materials like TiO_2 and ZnO under UV or solar irradiation can speed up plastic degradation. On the other hand, electrochemical oxidation methods such as Electro-Fenton (EFO) use Fe_2O_2 and H_2O_2 to create •OH, which oxidizes hazardous substances. High expenses and energy requirements continue to be challenges despite their potential (Jeong et al., 2023; Matavos-Aramyan, 2024).

4.4.3 Bioremediation and Phytoremediation

By using microorganisms, including bacteria, fungi, microalgae, and even insects, biodegradation turns plastics into non-toxic byproducts. Pseudomonas species and Bacillus cereus have both been shown to be effective at breaking down polyethylene and polypropylene (Jebashalomi et al., 2024; Xue et al., 2025). Enzymes such as laccase and cutinase play a vital role in depolymerizing plastics into smaller molecules that microbes can digest (Suresh et al., 2025; Temporiti et al., 2022).

In addition to microbiological processes, phytoremediation provides a low-cost, environmentally friendly solution by allowing plants to absorb, accumulate, or transform plastic pollutants. Certain species have the ability to absorb nanoplastics from soil and water through their root systems. While roots are the most affected plant organs, nanoplastic absorption and transfer can cause oxidative stress and reduced photosynthesis in plants. According to research, polystyrene particles can accumulate up in wheat and lettuce, which can lead to growth retardation and changed nutritional absorption (Gao et al., 2021; Jadhav & Medyńska-Juraszek, 2024).

4.4.4 Emerging Approaches

According to recent research, integrated remediation techniques combine biological degradation with chemical pre-treatment (such as UV irradiation). The mechanism of photodegradation is shown in Fig. 4.2. To improve degradation rates, for example, pre-exposing plastics to UV light increases their vulnerability to microbial enzymatic attack (Crystal Thew et al., 2024; Mohanan et al., 2020; Qutob et al., 2025). Combining photocatalysis and enzymatic degradation holds potential for boosting productivity while using less energy (Maciá-Agulló et al., 2015).

Adsorbing nanoplastics from aqueous media has demonstrated over 99% efficiency for biochars enabled by nanotechnology, such as biochar made from sugarcane bagasse (Ganie et al., 2021). These materials serve as both catalytic platforms and adsorbents, providing multipurpose solutions.

4.5 Circular Economy of Plastics

The concept of a circular economy (CE) for plastics arises as a transformational solution to reduce the environmental impact of plastic waste and shift away from the existing linear "take-make-dispose" model (Harun-Ur-Rashid, 2025). The goals of a circular economy are to maximize the value of resources while they are in use, recover and regenerate products at the end of their useful lives, and keep materials in use for as long as possible.

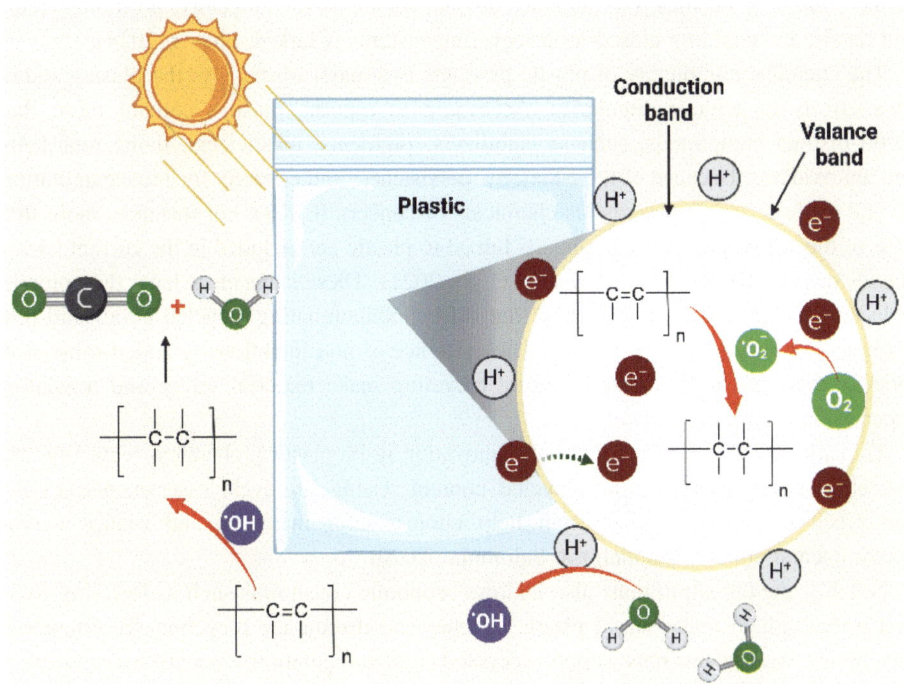

Fig. 4.2 Photodegradation mechanism of plastic (*Created by using Biorender.com*)

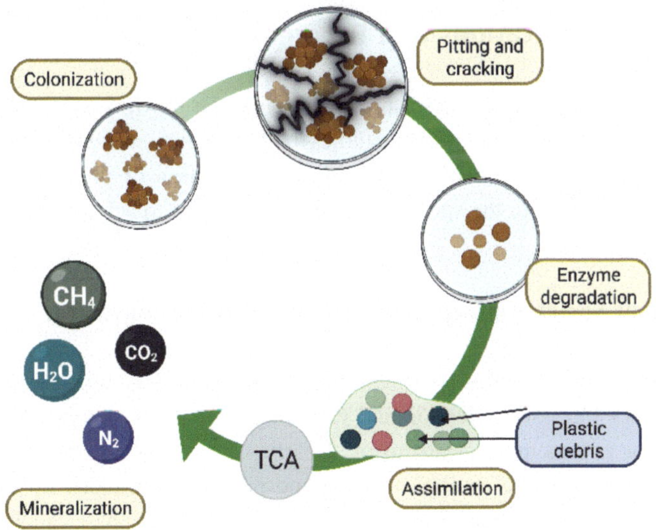

Fig. 4.3 Microbial degradation of plastic (*Created by using Biorender.com*)

In the context of plastics, this includes developing items for longevity, promoting reuse and repair, and enabling closed-loop recycling systems (Clark & Shaver, 2024).

The chemical complexity of plastic products is a major obstacle to the plastics industry's efforts to achieve circularity. According to reports, plastics contain more than 6,000 distinct compounds, such as stabilizers, pigments, plasticizers, flame retardants, and antioxidants. Because of their toxicity, persistence, and capacity for bioaccumulation, several of these are categorized as chemicals of concern (CoCs). For instance, more than 25% of the 1,518 high-risk compounds linked to plastic are included in the current hazard classes or prioritization lists (Aurisano et al., 2021). These substances have the potential to leach out during the use and recycling stages, contaminating recycled goods and posing risks to humans and ecosystems. The existence of non-intentionally added substances (NIAS) and cross-contamination during recycling make material safety and regulatory supervision even more difficult.

Transitioning to circularity requires innovation in bioplastics, chemical recycling, and traceable supply networks for recycled content. Using life-cycle assessments (LCAs) and green chemistry concepts can help choose safer materials and reduce overall environmental effects (Mondello & Salomone, 2020).

Notably, the transition must also address economic constraints such as lowering fossil fuel prices, which make virgin plastic cheaper and discourage recycling. To counteract this, policy mechanisms must support recycled content regulations, incentivize eco-design, and promote eco-labelling schemes to educate customers and drive market demand for sustainable products. Figure 4.4 shows the circular economy of plastic waste.

4.6 Policy Implications and Recommendations

Effective policymaking remains critical to addressing the growing issue of plastic waste. Global experiences and national case studies both demonstrate that no one policy can completely reduce plastic pollution; instead, a combination of policies that work well together and are tailored to the specific situation is required to achieve significant results.

4.6.1 Policy Shortcomings and the Need for Integration

India's current plastic waste management framework is widely acknowledged as inefficient, with issues like low collection and recycling rates, high volumes of unsegregated waste, and poor enforcement mechanisms. Nearly 40% of India's plastic waste remains unmanaged, contributing to ecological harm and public health risks. In global contexts as well, policies tend to focus heavily on the consumption phase, such as bans and

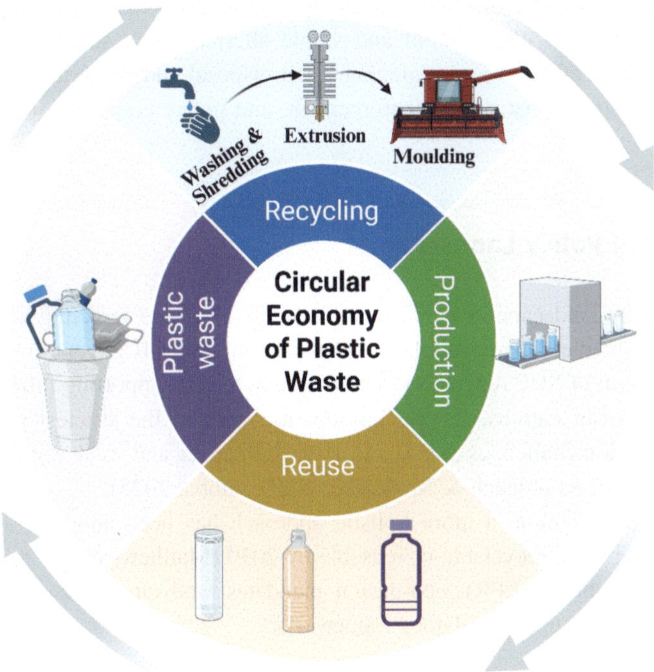

Fig. 4.4 Circular economy of plastic waste (*Created by using Biorender.com*)

user charges for single-use plastics, while production and disposal phases remain under-regulated, creating a gap in comprehensive life-cycle management (Hunt, 1996; Kumar et al., 2017).

4.6.2 System Dynamics Modeling

System Dynamics Modeling (SDM) is a computational approach used to simulate the behavior of complex systems over time (Zhao; MacLeod et al., 2023). It considers the interconnections, feedback loops, and time delays that exist between various system components. In the context of plastic waste, this means modeling the entire life cycle—from plastic production and consumption to disposal, recycling, and leakage into the environment—and simulating the impact of different policy scenarios. Researchers developed a dynamic simulation model to evaluate the outcomes of various policy combinations on plastic waste generation, recycling rates, and environmental leakage (Dhanshyam & Srivastava, 2021). As applied in the Indian context, it shows that phased implementation of a composite policy mix, including kerbside recycling facilities, disposal fees, and recycling subsidies, has the greatest impact on lowering plastic waste stock. For instance, disposal fees and a phased rollout of kerbside recycling facilities worked better together than they did separately. Without enforcement and viable alternatives, a stand-alone plastic ban might have adverse effects by encouraging illicit disposal and black-market manufacturing. This demonstrates that timing, enforcement, and policy combinations are critical to success.

4.6.3 Global Policy Landscape

The global policy landscape reveals a dominance of measures targeting single-use plastics (SUPs), particularly bans on plastic bags and microbeads. Over 60 countries have enacted some form of SUP regulation, but enforcement and supporting infrastructure vary significantly. African nations, despite introducing some of the strictest penalties, often face implementation challenges due to a lack of alternatives and weak regulatory capacity (Adam et al., 2020; Knoblauch & Mederake, 2021; Jumba, 2025).

In the European Union, a more holistic approach has been adopted, aiming for all plastic packaging to be recyclable or reusable by 2030 (Matthews et al., 2021). Extended Producer Responsibility (EPR), eco-design mandates, and circular economy strategies form the backbone of their regulatory frameworks.

4.6.4 Policy Implications and Recommendations

4.6.4.1 Lifecycle-Based Regulatory Framework

According to a lifecycle-based perspective, plastic waste starts from the design stage and continues during usage, disposal, and after consumption. The focus of current laws, such as plastic bans, is frequently on consumption and disposal, with little attention paid to upstream interventions like green sourcing, product design, and material innovation (Alaghemandi, 2024; Williams & Rangel-Buitrago, 2022).

Eco-design requirements demand that producers design items that are composed of monomaterials, are easily recyclable, and don't contain any hazardous additives. Additionally, they also promote bio-based and biodegradable substitutes for some materials (like food packaging), especially in situations where recycling is not practical (Donkor et al., 2023; Ries et al., 2023). To facilitate segregation and traceability, plastics should have uniform labels that specify the kind of polymer, whether they are recyclable, and their chemical composition.

It is recommended that national policy tools incorporate life-cycle assessments (LCAs) to examine the environmental impact of plastics at all stages of manufacture, usage, and end-of-life.

4.6.4.2 Reinforced Extended Producer Responsibility (EPR)

With EPR, producers and brand owners bear the responsibility for managing plastic trash instead of consumers and municipalities. Despite being included in several national frameworks (for example, India's PWM Rules 2016), EPR is frequently poorly enforced (Aurisano et al., 2021; Jumba, 2025).

To ensure transparency and accountability, digital tracking systems require producers to disclose the amount of plastic they introduce and remove from the market (Rumetshofer & Fischer, 2023). Digital EPR websites, such as the CPCB portal in India, can monitor compliance. Reduced rates for eco-designed items will encourage recyclability, whereas higher EPR fees should be applied to plastics that are difficult to recycle (such as multilayer packaging). To operationalize EPR more effectively, certified industry responsibility organizations (PROs) should be established to oversee data collection, processing, and reporting (Tumu et al., 2023). These should be transparent and involve the informal sector. Performance-based incentives will further strengthen EPR by giving manufacturers subsidies if they use a lot of post-consumer recycling, invest in R&D, or surpass collection and recycling goals.

4.6.4.3 Localized Implementation and Infrastructure Support

The success of policies depends on local ability and implementation mechanisms in addition to national frameworks. Local governments in many developing nations lack the funding necessary to implement laws, oversee collection networks, or construct facilities for recycling and segregation (Dagilienė et al., 2021). Grants for infrastructure such as collection vehicles, MRFs (Material Recovery Facilities), and transfer stations might be allocated from the municipal budget. Training programs on circular waste management will increase the knowledge and build the capacity among sanitation staff, urban planners, and municipal engineers to handle waste (Herat, 2015). Helping local NGOs or social enterprises to establish low-tech recycling facilities in rural locations. Encourage zero-waste villages, ward-level segregation units, and cooperatives of waste pickers—integrating them formally into the waste economy.

This inclusive and ground-level approach ensures community ownership, job creation, and higher levels of compliance with national regulations.

4.6.4.4 Circular Economy Incentives and Green Finance

Economic instruments are key to accelerating the transition to a circular plastic economy. These resources encourage private investment in the waste industry and increase the competitiveness of recycled plastics relative to virgin materials.

Tools and mechanisms:

- Plastic taxes: To absorb the expense of pollution, environmental fees should be imposed on virgin plastic.

- Mandates for recycled content: Demand that producers use a minimum proportion of recycled polymers in industrial products and packaging.
- Startup incubation: Provide grants, pilot funding, and market access to entrepreneurs developing bioplastics, chemical recycling, and plastic-to-fuel technologies.

These tools make sustainable materials and services economically viable, paving the way for a circular ecosystem driven by market forces and not just mandates (Agrawal et al., 2024; Kumar et al., 2023).

4.6.4.5 Encourage Industry Responsibility and Corporate Accountability

Corporate Social Responsibility (CSR) can play a key role in reducing plastic waste by encouraging businesses to take meaningful action. Companies can demonstrate their commitment to sustainability by minimizing the use of plastic packaging, exploring eco-friendly alternatives, and supporting circular economy initiatives. Through CSR programs, businesses can not only reduce their environmental footprint but also build stronger connections with environmentally conscious consumers. Such initiatives may involve product redesign using recyclable components, financial support for environmental clean-up efforts, and collaborations with waste reduction organizations. By actively participating in such initiatives, the private sector can help drive long-term change in how plastic is produced, used, and managed (Blasiak et al., 2021; Bokor, 2024).

4.7 Conclusion

The environmental challenges posed by plastic pollution are vast, interlinked, and urgent. As explored throughout this chapter, addressing these challenges requires a multidimensional and future-focused approach that integrates science, innovation, governance, and community action. The rise in plastic production and its persistence in ecosystems have pushed conventional management systems beyond their capacity, necessitating the adoption of transformative strategies. A critical first step lies in improving how we measure and monitor plastic pollution, particularly microplastics, which now permeate soils, oceans, and even food systems. Accurate data enables targeted interventions and informs public health risk assessments and policy frameworks. Meanwhile, innovations in recycling technologies—from advanced mechanical and chemical methods to emerging biological pathways—present a growing arsenal of tools that can recover material value while minimizing environmental harm.

Simultaneously, the chapter demonstrates how remediation strategies, including bioremediation, advanced oxidation processes, and phytoremediation, are evolving to remove or neutralize existing plastic contaminants in various ecosystems. These solutions, especially when integrated, offer scalable potential for both cleanup and long-term ecological restoration.

Transitioning toward a circular plastic economy remains central to these efforts. This requires not just closing material loops through recycling and reuse but also eliminating harmful chemical additives, enhancing product design, and enabling safe, traceable, and high-quality recycled content. Circularity must be driven not only by innovation but also by supportive policy mechanisms, economic instruments, and strong enforcement.

The final section of the chapter emphasizes that policy design must be systemic, sequenced, and inclusive—targeting all stages of the plastic life cycle. From lifecycle-based regulations and EPR to localized infrastructure support and corporate accountability, policy must operate hand-in-hand with technological and societal progress. System dynamics modeling reinforces this by showing how combinations of policies, implemented in the right order, lead to the greatest long-term impact.

References

Achilias, D. S. (2025). Thermo-chemical recycling of plastics as a sustainable approach to the plastic waste issue. *Euro-Mediterranean Journal for Environmental Integration, 2025*, 1–14. https://doi.org/10.1007/S41207-025-00800-7

Adam, I., Walker, T. R., Bezerra, J. C., & Clayton, A. (2020). Policies to reduce single-use plastic marine pollution in west Africa. *Marine Policy, 116*, Article 103928. https://doi.org/10.1016/J.MARPOL.2020.103928

Agrawal, R., Agrawal, S., Samadhiya, A., et al. (2024). Adoption of green finance and green innovation for achieving circularity: An exploratory review and future directions. *Geoscience Frontiers, 15*, Article 101669. https://doi.org/10.1016/J.GSF.2023.101669

Alaghemandi, M. (2024). Sustainable solutions through innovative plastic waste recycling technologies. *Sustainability (Switzerland), 16*.

Aurisano, N., Weber, R., & Fantke, P. (2021). Enabling a circular economy for chemicals in plastics. *Current Opinion in Green and Sustainable Chemistry, 31*, Article 100513. https://doi.org/10.1016/J.COGSC.2021.100513

Barceló, D., Picó, Y., & Alfarhan, A. H. (2023). Microplastics: Detection in human samples, cell line studies, and health impacts. *Environmental Toxicology and Pharmacology, 101*, Article 104204. https://doi.org/10.1016/J.ETAP.2023.104204

Bhavsar, P., Bhave, M., & Webb, H. K. (2023). Solving the plastic dilemma: The fungal and bacterial biodegradability of polyurethanes. *World Journal of Microbiology & Biotechnology, 39*, 1–11. https://doi.org/10.1007/S11274-023-03558-8/TABLES/2

Blasiak, R., Leander, E., Jouffray, J. B., & Virdin, J. (2021). Corporations and plastic pollution: Trends in reporting. *Sustainable Futures, 3*, Article 100061. https://doi.org/10.1016/J.SFTR.2021.100061

Bokor, B. (2024). Corporate engagement in mitigating plastic pollution: Examining voluntary initiatives and EU regulations. *Frontiers in Sustainability, 5*, 1420041. https://doi.org/10.3389/FRSUS.2024.1420041/BIBTEX

Campanale, C., Savino, I., Massarelli, C., & Uricchio, V. F. (2023). Fourier transform infrared spectroscopy to assess the degree of alteration of artificially aged and environmentally weathered microplastics. *Polymers (Basel), 15*, 911. https://doi.org/10.3390/POLYM15040911

Clark, R. A., & Shaver, M. P. (2024). Depolymerization within a circular plastics system. *Chemical Reviews, 124*, 2617–2650.

Crystal Thew, X. E., Lo, S. C., Ramanan, R. N., et al. (2024). Enhancing plastic biodegradation process: Strategies and opportunities. *Critical Reviews in Biotechnology, 44*, 477–494. https://doi.org/10.1080/07388551.2023.2170861;SUBPAGE:STRING:FULL

Dagilienė, L., Varaniūtė, V., & Bruneckienė, J. (2021). Local governments' perspective on implementing the circular economy: A framework for future solutions. *Journal of Cleaner Production, 310*, Article 127340. https://doi.org/10.1016/J.JCLEPRO.2021.127340

Dhanshyam, M., & Srivastava, S. K. (2021). Effective policy mix for plastic waste mitigation in India using System Dynamics. *Resources, Conservation and Recycling, 168*, Article 105455. https://doi.org/10.1016/J.RESCONREC.2021.105455

Donkor, L., Kontoh, G., Yaya, A., et al. (2023). Bio-based and sustainable food packaging systems: Relevance, challenges, and prospects. *Applied Food Research, 3*, Article 100356. https://doi.org/10.1016/J.AFRES.2023.100356

Dorigato, A. (2021). Recycling of polymer blends. *Advanced Industrial and Engineering Polymer Research, 4*, 53–69. https://doi.org/10.1016/J.AIEPR.2021.02.005

Ganie, Z. A., Khandelwal, N., Tiwari, E., et al. (2021). Biochar-facilitated remediation of nanoplastic contaminated water: Effect of pyrolysis temperature induced surface modifications. *Journal of Hazardous Materials, 417*, Article 126096. https://doi.org/10.1016/J.JHAZMAT.2021.126096

Gao, M., Xu, Y., Liu, Y., et al. (2021). Effect of polystyrene on di-butyl phthalate (DBP) bioavailability and DBP-induced phytotoxicity in lettuce. *Environmental Pollution, 268*, Article 115870. https://doi.org/10.1016/J.ENVPOL.2020.115870

Gasde, J., Woidasky, J., Moesslein, J., & Lang-Koetz, C. (2020). Plastics recycling with tracer-based-sorting: Challenges of a potential radical technology. *Sustainability, 13*, 258. https://doi.org/10.3390/SU13010258

Harun-Ur-Rashid, M. (2025). Circular economy and depolymerization: Creating value from waste. *ACS Symposium Series*, 169–193. https://doi.org/10.1021/BK-2025-1501.CH008

Herat, S. (2015). Waste management training and capacity building for local authorities in developing countries. *Waste Management and Research, 33*, 1–2. https://doi.org/10.1177/0734242X14565543/ASSET/E82F96DE-3FD3-4943-B410-02071E819431/ASSETS/IMAGES/LARGE/10.1177_0734242X14565543-IMG1.JPG

Hunt, C. (1996). Child waste pickers in India: The occupation and its health risks. *Environment and Urbanization, 8*, 111–114. https://doi.org/10.1177/095624789600800209

Ikegwu, U. M., Munguía-López, A. D. C., Zavala, V. M., & Van Lehn, R. C. (2025) Screening green solvents for multilayer plastic film recycling processes. *Computers & Chemical Engineering, 199*, 109129. https://doi.org/10.1016/J.COMPCHEMENG.2025.109129

Jadhav, B., & Medyńska-Juraszek, A. (2024). Microplastic and nanoplastic in crops: Possible adverse effects to crop production and contaminant transfer in the food chain. *Plants, 13*, 2526. https://doi.org/10.3390/PLANTS13172526

Jebashalomi, V., Emmanuel Charles, P., & Rajaram, R. (2024). Microbial degradation of low-density polyethylene (LDPE) and polystyrene using Bacillus cereus (OR268710) isolated from plastic-polluted tropical coastal environment. *Science of the Total Environment, 924*, Article 171580. https://doi.org/10.1016/J.SCITOTENV.2024.171580

Jeong, Y., Gong, G., Lee, H. J., et al. (2023). Transformation of microplastics by oxidative water and wastewater treatment processes: A critical review. *Journal of Hazardous Materials, 443*, Article 130313. https://doi.org/10.1016/J.JHAZMAT.2022.130313

Joho, Y., Vongsouthi, V., Gomez, C., et al. (2024). Improving plastic degrading enzymes via directed evolution. *Protein Engineering, Design and Selection, 37*, gzae009. https://doi.org/10.1093/PROTEIN/GZAE009

Jumba, K. K. (2025). Policy frameworks and governance in plastic waste management: A comparative analysis of African countries. *Research Invention Journal Of Engineering And Physical Sciences, 4*, 13–17. https://doi.org/10.59298/RIJEP/2025/411317

Jung, E. S., Choe, J. H., Kim, J. S., et al. (2024). Quantitative Raman analysis of microplastics in water using peak area ratios for concentration determination. *NPJ Clean Water, 7*, 1–6. https://doi.org/10.1038/S41545-024-00397-4; SUBJMETA = 169,172,2805,704,706; KWRD = ENVIRONMENTAL + CHEMISTRY, WATER + RESOURCES

Knoblauch, D., & Mederake, L. (2021). Government policies combatting plastic pollution. *Current Opinion in Toxicology, 28*, 87–96. https://doi.org/10.1016/J.COTOX.2021.10.003

Kotar, S., McNeish, R., Murphy-Hagan, C., et al. (2022). Quantitative assessment of visual microscopy as a tool for microplastic research: Recommendations for improving methods and reporting. *Chemosphere, 308*, Article 136449. https://doi.org/10.1016/J.CHEMOSPHERE.2022.136449

Kumar, B., Kumar, L., Kumar, A., et al. (2023). Green finance in circular economy: A literature review. *Environment, Development and Sustainability, 26*(7), 16419–16459. https://doi.org/10.1007/S10668-023-03361-3

Kumar, S., Smith, S. R., Fowler, G., et al. (2017). Challenges and opportunities associated with waste management in India. *Royal Society Open Science, 4*. https://doi.org/10.1098/RSOS.160764/SUPPL_FILE/RSOS160764_REVIEW_HISTORY.PDF

Kwon, J., Kim, H., Siddiqui, M. Z., et al. (2025). A comprehensive pyrolysis-gas chromatography/mass spectrometry analysis for the assessment of microplastics in various salts. *Food Chemistry, 467*, Article 142193. https://doi.org/10.1016/J.FOODCHEM.2024.142193

Lange, J. P. (2021). Managing plastic waste-sorting, recycling, disposal, and product redesign. *ACS Sustainable Chemistry & Engineering, 9*, 15722–15738.

Leone, G., Moulaert, I., Devriese, L. I., et al. (2023). A comprehensive assessment of plastic remediation technologies. *Environment International, 173*, Article 107854. https://doi.org/10.1016/J.ENVINT.2023.107854

Maciá-Agulló, J. A., Corma, A., & Garcia, H. (2015). Photobiocatalysis: The power of combining photocatalysis and enzymes. *Chemistry—A European Journal, 21*, 10940–10959. https://doi.org/10.1002/CHEM.201406437;WEBSITE:WEBSITE:CHEMISTRY-EUROPE;REQUESTEDJOURNAL:JOURNAL:15213765;WGROUP:STRING:PUBLICATION

MacLeod, M., Domercq, P., Harrison, S., & Praetorius, A. (2023). Computational models to confront the complex pollution footprint of plastic in the environment. *Nature Computational Science, 3*, 486–494. https://doi.org/10.1038/S43588-023-00445-Y

Mariano, S., Tacconi, S., Fidaleo, M., et al. (2021). Micro and nanoplastics identification: Classic methods and innovative detection techniques. *Frontiers in Toxicology, 3*, Article 636640. https://doi.org/10.3389/FTOX.2021.636640

Matavos-Aramyan, S. (2024). Addressing the microplastic crisis: A multifaceted approach to removal and regulation. *Environmental Advances, 17*, Article 100579. https://doi.org/10.1016/J.ENVADV.2024.100579

Matthews, C., Moran, F., & Jaiswal, A. K. (2021). A review on European Union's strategy for plastics in a circular economy and its impact on food safety. *Journal of Cleaner Production, 283*, Article 125263. https://doi.org/10.1016/J.JCLEPRO.2020.125263

McNeeley, A., & Liu, Y. A. (2024). Assessment of PET depolymerization processes for circular economy. 2. Process design options and process modeling evaluation for methanolysis, glycolysis, and hydrolysis. *Industrial and Engineering Chemistry Research, 63*, 3400–3424. https://doi.org/10.1021/ACS.IECR.3C04001/SUPPL_FILE/IE3C04001_SI_001.PDF

Mohanan, N., Montazer, Z., Sharma, P. K., & Levin, D. B. (2020). Microbial and enzymatic degradation of synthetic plastics. *Frontiers in Microbiology, 11*, Article 580709. https://doi.org/10.3389/FMICB.2020.580709

Mondello, G., & Salomone, R. (2020). Assessing green processes through life cycle assessment and other LCA-related methods. In *Studies in surface science and catalysis* (pp. 159–185). Elsevier.

Mong, G. R., Chong, C. T., Chong, W. W. F., et al. (2022). Progress and challenges in sustainable pyrolysis technology: Reactors, feedstocks and products. *Fuel, 324*, Article 124777. https://doi.org/10.1016/J.FUEL.2022.124777

Olawade, D. B., Fapohunda, O., Wada, O. Z., et al. (2024). Smart waste management: A paradigm shift enabled by artificial intelligence. *Waste Management Bulletin, 2*, 244–263. https://doi.org/10.1016/J.WMB.2024.05.001

Parga Martínez, K. B., da Silva, V. H., Andersen, T. J., et al. (2023). Improved separation and quantification method for microplastic analysis in sediment: A fine-grained matrix from Arctic Greenland. *Marine Pollution Bulletin, 196*, Article 115574. https://doi.org/10.1016/J.MARPOLBUL.2023.115574

Park, W. J., Hwangbo, M., & Chu, K. H. (2023). Plastisphere and microorganisms involved in polyurethane biodegradation. *Science of the Total Environment, 886*, Article 163932. https://doi.org/10.1016/J.SCITOTENV.2023.163932

Prata, J. C., da Costa, J. P., Duarte, A. C., & Rocha-Santos, T. (2019). Methods for sampling and detection of microplastics in water and sediment: A critical review. *TrAC Trends in Analytical Chemistry, 110*, 150–159. https://doi.org/10.1016/J.TRAC.2018.10.029

Qutob, M., Rafatullah, M., & Flafel, H. M. (2025). Advancement into the integration of chemical oxidation and bioremediation for the removal of hydrocarbon-contaminated soil. *Journal of Environmental Chemical Engineering, 13*, Article 116846. https://doi.org/10.1016/J.JECE.2025.116846

Ries, L., Wartzack, S., & Zipse, O. (2023). Creating sustainable products: The road to circularity. *Road to Net Zero: Strategic Pathways for Sustainability-Driven Business Transformation*, 123–157. https://doi.org/10.1007/978-3-031-42224-9_5/FIGURES/1

Rumetshofer, T., & Fischer, J. (2023). Information-based plastic material tracking for circular economy—a review. *Polymers, 15*, 1623. https://doi.org/10.3390/POLYM15071623

Saebea, D., Ruengrit, P., Arpornwichanop, A., & Patcharavorachot, Y. (2020). Gasification of plastic waste for synthesis gas production. *Energy Reports, 6*, 202–207. https://doi.org/10.1016/J.EGYR.2019.08.043

Sarkar, S., Diab, H., & Thompson, J. (2023). Microplastic pollution: Chemical characterization and impact on wildlife. *International Journal of Environmental Research and Public Health, 20*, 1745. https://doi.org/10.3390/IJERPH20031745

Schrank, I., Möller, J. N., Imhof, H. K., et al. (2022). Microplastic sample purification methods—assessing detrimental effects of purification procedures on specific plastic types. *Science of the Total Environment, 833*, Article 154824. https://doi.org/10.1016/J.SCITOTENV.2022.154824

Sevilla, M. E., Garcia, M. D., Perez-Castillo, Y., et al. (2023). Degradation of PET Bottles by an Engineered Ideonella sakaiensis PETase. *Polymers (Basel), 15*, 1779. https://doi.org/10.3390/POLYM15071779/S1

Sharma, P., Sharma, P., & Abhishek, K. (2024). Sampling, separation, and characterization methodology for quantification of microplastic from the environment. *Journal of Hazardous Materials Advances, 14*, Article 100416. https://doi.org/10.1016/J.HAZADV.2024.100416

Singh Jadaun, J., Bansal, S., Sonthalia, A., et al. (2022). Biodegradation of plastics for sustainable environment. *Bioresource Technology, 347*, Article 126697. https://doi.org/10.1016/j.biortech.2022.126697

Suresh, V., Shams, R., Dash, K. K., et al. (2025). Comprehensive review on enzymatic polymer degradation: A sustainable solution for plastics. *J Agric Food Res, 20*, Article 101788. https://doi.org/10.1016/J.JAFR.2025.101788

Tasseron, P. F., van Emmerik, T. H. M., de Winter, W., et al. (2024). Riverbank plastic distributions and how to sample them. *Microplastics and Nanoplastics, 4*, 1–12. https://doi.org/10.1186/S43591-024-00100-X/FIGURES/6

Temporiti, M. E. E., Nicola, L., Nielsen, E., & Tosi, S. (2022). Fungal enzymes involved in plastics biodegradation. *Microorganisms, 10*, 1180. https://doi.org/10.3390/MICROORGANISMS10061180

Tumu, K., Vorst, K., & Curtzwiler, G. (2023). Global plastic waste recycling and extended producer responsibility laws. *Journal of Environmental Management, 348*, Article 119242. https://doi.org/10.1016/J.JENVMAN.2023.119242

Umdagas, L., Orozco, R., Heeley, K., et al. (2025). Advances in chemical recycling of polyethylene terephthalate (PET) via hydrolysis: A comprehensive review. *Polymer Degradation and Stability, 234*, Article 111246. https://doi.org/10.1016/J.POLYMDEGRADSTAB.2025.111246

Wang, S., Jian, Q., Zhang, P., et al. (2022). Tracing land-based microplastic sources in coastal waters of Zhanjiang Bay, China: Spatiotemporal pattern, composition, and flux. *Frontiers in Marine Science*, 9. https://doi.org/10.3389/fmars.2022.934707

Williams, A. T., & Rangel-Buitrago, N. (2022). The past, present, and future of plastic pollution. *Marine Pollution Bulletin, 176*, Article 113429. https://doi.org/10.1016/J.MARPOLBUL.2022.113429

Xue, H., Chen, X., Jiang, Z., et al. (2025). Biodegradation of polypropylene by Bacillus cereus PP-5 isolated from waste landfill. *Ecotoxicology and Environmental Safety, 296*, Article 118205. https://doi.org/10.1016/J.ECOENV.2025.118205

Zhao, X. *Pollution modeling plastic pollution: A system dynamics approach to understand and control macroscopic plastic flow dynamics.*

The manufacturer's authorised representative in the EU is Springer Nature Customer Service Centre GmbH, Europaplatz 3, 69115 Heidelberg, Germany. If you have any concerns regarding our products, please contact ProductSafety@springernature.com

Printed and bound by CPI Group (UK) Ltd, Croydon, CR0 4YY
15/07/2026
02167645-0014